THE CITY

AMERICAN CITY POLITICS

POLITICS OF THE CITY

A note on terms

There are a number of terms which students will know or quickly pick up; but one or two may lead to confusion.

Charter: constitution, or basic law.

Planning: connotes comprehensive planning of total community services, not just the physical layout.

Zoning: the regulation of land-use and development (thus planning in the narrow sense).

Merit or civil service systems: appointment and promotion by formal procedures based on competitive assessment of professional competence, usually by examination.

Patronage: power to award appointments, contracts and other favours; used (i) to secure politically compatible policy advisors and high executives (ii) to reward party workers (iii) to secure support (iv) for personal profit. While the whole system is open to corruption, the first three objectives are accepted elements in American politics.

Administration: may refer, according to context, to the Government (especially the Executive) as in 'the Nixon Administration'; and is not restricted to 'civil servants'.

Official, professional: are used about elected officers like the Mayor, and not confined to career 'civil servants'.

Labour: trade unions.

Suburbs: places regarded as suburbs are often farther out of the city centre than in Britain.

Middle class: as a general reference is more inclusive than in Britain: if you are not middle class you are poor.

Metropolitan: see p. 2.

Machine: see p. 8.

Reform: see p. 9.

Acknowledgements

I am convinced of the pedagogic justification for small books on large subjects. At the same time I am aware that such books depend on the work of many others, who have laboured at making large books on smaller subjects. So my first and deepest debt is to those American scholars whose work I have drawn on. The names scattered through the text indicate my greatest creditors. Those who pursue these references will discover much fine scholarship, graced by some notable examples of elegant, forceful and readable prose.

I am also indebted to several administrators in Washington DC and New York for making so welcome a wandering British academic. I am grateful to Miss Judy Morton, formerly a Research Officer in this Department, for allowing me to consult some of her own work on the City Manager, and to Miss Sandra Winterhalder, a graduate research student, for reading the manuscript, and pointing out some of its more remediable defects.

Finally, I must offer my thanks to my wife for a good deal of typing and even more forbearance.

P.J.M.

Contents

First published 1970
by Routledge & Kegan Paul Ltd
Broadway House, 68-74 Carter Lane
London E.C.4
Printed in Great Britain
by Northumberland Press Ltd
Gateshead

ISBN 0 7100 6807 7

American City Politics

by P. J. Madgwick

Lecturer in Political Science,
University College of Wales, Aberystwyth

LONDON
ROUTLEDGE & KEGAN PAUL

First published in 1970

This edition published in 2007
Routledge
2 Park Square, Milton Park, Abingdon, Oxfordshire OX14 4RN
711 Third Avenue, New York, NY 10017

First issued in paperback 2014

Routledge is an imprint of Taylor & Francis Group, an informa business

British Library Cataloguing in Publication Data
A CIP catalogue record for this book
is available from the British Library

American City Politics

ISBN13: 978-0-415-41771-6 (volume)
ISBN13: 978-0-415-75962-5 (pbk)

Routledge Library Editions: The City

AMERICAN CITY POLITICS

P.J. MADGWICK

Routledge
Taylor & Francis Group
LONDON AND NEW YORK

1

The city in the USA

The study of American city politics

The attraction of American local politics for the political scientist is that there is so much of it. For the local government of 200 million people spread over half a continent in 90,000 legally separate units produces a fascinating variety of types of governmental structure and political processes. The difficulties of making general statements of any validity and significance about such a multiplicity of activity are formidable. To forsake the analysis of a single tree for the description of the wood is hazardous indeed, not least because it assumes that the trees make up a wood. Conceivably the totality of activity is as diverse as say, an amalgamation of a cricket match, a baseball game, an investiture and a space-shot. It is true that no city or group of cities may stand alone as a representative example of American city government. 'No one can ever judge the quality of local government in the United States by his experience with one or two units' (Dahl, 1967, 180). However the limits to the scope of generalization are not entirely disabling.

Fortunately, the literature, though biased towards the North Eastern United States, is wide ranging; and a judgement of quality (as distinct from an understanding of structure and process) is not a dominant purpose in this study. Nor is the national reference of prime importance. The object is not so much to encapsulate *American* local government, as to illumine kinds of local political activity *in America*. The concern is as much with parts of the wood, and sometimes with individual trees, as with the whole amorphous and cacophonous wood. Hence, the use of 'some', 'most', 'sometimes', 'usually' and similar approximations may perhaps be justified where approximation is illuminating, exactness impossible and silence pedagogically useless (or frustrating anyway).

The term 'city' itself is the first approximation. In the USA it has precise meaning only in constitutional law, indicating a particular division of local government. Here it is used in the imprecise but generally accepted sense of an urban as distinct from a rural area, characterized by large numbers of people, living in proximity, mainly engaged in industrial and commercial activity and interacting socially in some ways so as to form a large, if rather tenuous and shadowy, community. (But that word is an approximation, too.) Most such areas are legally cities, but stray into the surrounding area, like British cities. When this results in a cluster of population around, and related to, a city of at least 50,000 inhabitants, the area is officially described as a standard metropolitan statistical area, or a metropolis, or in planners' polemical, the Spread City. In 1960 there were 212 SMSA's containing 63% of the population of the USA. The smallest city mentioned in the text is Oberlin (Ohio) with 8,000 people; the largest is New York City (with 8,000,000). Most of the cities considered here are larger than 50,000, and there are many references to the very largest cities of over half a million inhabitants. This is not a disproportionate emphasis in relation to the number of people living in these cities; but it is necessary to bear in mind the large number of smaller cities, and the high standing of small cities in the American ethos.

Cities have always had an important place in American social thought (see Glaab and Brown, 1967). In the nineteenth century, when New Paris, Rome City and the like were being deliberately constructed in the wilderness, cities were the object of both sonorous vilification and extravagant 'boosting'. The city is now, as much as in the last century, in the forefront of much social thought. 'The city,' proclaimed Max Lerner, 'is the battleground of the values of the culture' (Lerner, 1958, 168). It is the theatre of the contemporary 'urban crisis'. There is indeed a new academic 'booster' movement promoting the city. It is conceivable that urban perspectives may be exploited as a facile Explanation or Key to History and Politics, as in their day were class conflict and the frontier. The study of urban politics is at best a partial view of politics of the American people.

The nature of the urban community

The pace of urbanization in the USA was dramatically fast, even by the standards of the British industrial revolution. In 1820 there were only twelve cities with populations larger than 10,000; by 1860 the figure was 101. Of these, eight had more than 100,000 inhabitants; New York with Brooklyn contained over a million people and was then the third city in the world. Between 1915 and 1920 the number of people living in 'towns' of more than 2,500 first exceeded 50% of the population and it is now above 70%. Urbanization at this pace, coupled with massive immigration, expansion westward to the Pacific, and rapid industrialization, represented a profound historical experience which has left its mark on American society and the American people.

For a historically conscious nation the legacy of urbanization has been rich but perplexing. It was a visible triumph of enterprise and technical inventiveness, appearing to vindicate the 'booster' spirit of the pioneers. But it was also restless and unplanned; cities still show the rawness of their making, both in their town landscapes and buildings, and in their politics. Second, the idea of the city never made a complete conquest of the American ethos. Cities offered opportunity, comfort, glamour (Lerner) but corruption too. Rural myths and values had a strong counter-attraction. The American people are incompletely and unwillingly an urban nation. Reluctant to fight on Lerner's cultural battleground, they flee in large numbers to the suburbs (outer suburbs in English terms) and voting with their wheels, promote the spread or scattered city. This, to some visionaries of the city, is an anti-city.

More than half the population lives in or around the larger cities (50,000 population and more). Size would seem to have some significance for the nature of society and of politics since it affects the economic and financial structure, the level and kind of needs and demands, and the potential response of the municipality. There is evidence that large cities have a more vital and diverse group life; and both organized and informal political activity is related to size (Lee, 1960, 148-51). Larger cities tend to be more developed i.e. committed to economic development and the provision of amenities. However, the relationship between size and political characteristics is complicated by other partially independent factors—rate of growth, existence of immigrant and middle-class elements in the population, tradition and ethos, the orientation of policy makers (see Eyestone and Eulau, 1968, 56-64, also Kessel, 1962, 615-20).

Moreover, it is not at all clear whether there is any

significant character pertaining to a particular range of size. Obviously there is a low point of population at which social relationships change from face-to-face to indirect, and cease to extend predominantly to the whole community. The American statistical custom of counting urban places of more than 2,500 people as cities is valid here. At some higher point the needs and responses of the urban place are transformed by size, and the city becomes a collection of communities (e.g. city centre, and suburb, but also commuters, golfers, teachers). Williams and Adrian propose an intermediate category for cities of 25,000 to 250,000 persons, which they say, lose the face-to-face style of small-town politics, but still have 'a central system of communication in political and social affairs that is often lacking in the very large population centres' (Williams and Adrian, 1963, 14).

The connection between location and politics is more obscure. Wolfinger and Field claimed to demonstrate that region was a more important predictor of governmental forms and output than ethos, ethnicity or socio-economic character (Wolfinger and Field, 1966). Their conclusions run counter to much recent work which has stressed the significance of the reform ethos (see Lineberry and Fowler, 1967). Research is as yet too much concentrated in the North-East for development of regional explanations of urban politics. But so far it has not been conclusively shown that such a distinctive region as the South has a distinctive form of local politics. It seems likely in any case that regional differences are declining.

Ethnicity is of major significance in urban politics. It is associated with socio-economic character, status, and success drives, and is therefore difficult to isolate analytically. Dahl has shown (1961, 60) how ethnicity modified socio-economic factors in determining support of the Democratic Party in New Haven. In New York 'class consciousness never appears so strongly as ethnic awareness' (Lowi, 1964, 34).

Ethnic awareness is intensified by colour, and coloured minorities are now a characteristic feature of Northern cities. 'Central cities are becoming more heavily Negro while the urban fringes around them remain almost entirely white. The proportion of Negroes in all central cities rose steadily from 12% in 1950, to 17% in 1960, to 20% in 1966' (Kerner Report, 1968, 243). Washington DC is now two thirds Negro and in the next 20-25 years at least ten of the fourteen largest cities will have Negro populations of 25-50%. Every aspect of government in these cities is now conditioned by the existence of the coloured minorities.

Comparatively, the foreign-born whites constitute a major but declining influence as such, and do not threaten stability. In 1960, 19% of the population of US were foreign-born. Some of these were Puerto Ricans or Mexicans. The classic European immigration of the nineteenth and early twentieth centuries ended by the 1920's. But the social and political significance of an immigrant descent continues. The political history of cities like Boston and New York is a saga of domestic nationalism, Italians struggling with the Irish and both with Jews. De Sapio's victory in New York, 1949, and Impelliteri's election as Mayor in 1951, marked the advance of the Italians. In some areas, small but cohesive ethnic groups promote a politics of neighbourhood and status. This was true of Poles and Dutch in two of the four cities studied by Williams and Adrian (1963). In 1969 Italians won both Republican and Democratic primaries in New York City against Mayor Lindsay, preeminently a white Anglo-Saxon Protestant (WASP). If ethnic political drives have diminished, they still have considerable force and are served with respect by politicians whose own ethnic drives may not be strong. Religion reinforces ethnic differences. Nominations for elective office and appointments are still balanced, or at least made with some consideration of race and religion. Thus the political processes, contrary to the theory of the melting pot, help to maintain an awareness of differences.

Population movement within metropolitan areas has expressed and emphasized the conflicts of classes, race and colour. The suburbs are growing rapidly. Over 'three quarters of the white population increase, 1950-66, of 36 million occurred in the suburbs. These were 95% white and became 96% white by 1966. In the same years, 1950-66, 86% of the increase in Negro population took place within the central cities' (Kerner Report, 1968, 243).

Suburban America was mainly middle-class America—zoning and building regulations, and pressures on and from builders and bankers, have preserved suburbs as havens for the prosperous white. The commuting businessman kept a tenuous link with the City centre, but increasingly he has found his social and political life, and eventually, his business, in the suburb.

The political culture of the city

Most Americans believe that local government has at least some effect on their day-to-day lives—88% of respondents compared with 74% in Britain and West Germany (Almond and Verba, 1963, 81). A high proportion again believed they could do something about an unjust local regulation—77% in USA, 78% in Britain, 62% in Germany (p. 185). A comparatively high proportion favour community activity— 51% in USA to 39% in Britain, and a mere 22% in Germany (p. 169). Not unexpectedly, more people claim to understand local issues moderately well or very well, than national or international issues (Almond and Verba, survey quoted in Dahl, 1967, 201). In local as in national government there appears to be a prevailing sense of competence.

However, the sense of competence is not related to high levels of direct political participation, as indicated by voting turnout. This is on average about 40% in local elections, between 10% and 30% less in particular cities than the turnout for Presidential elections (Banfield and Wilson, 1963, 225). Indeed, the general sense of competence and favourable attitudes to participation hide pockets of cynicism and alienation. Nor are they incompatible with quite widespread negative attitudes to the scope and activity of local government. The competent citizen, acting from a sense of satisfaction or of hostility, may choose to participate only in order to reduce the municipality to a conservative, caretaker role. In general this attitude is found among lower income groups, who are comparatively unaware of social needs and of the capacity of urban governments to fulfill them. The better-off groups are more ready to promote civic amenities and prosperity (Williams and Adrian, 1963). Thus, in sum, the sense of competence in local government is associated with a marked lack of enthusiasm for the actual operations of local government.

There have been two major traditions in urban politics, the 'machine' and 'reform'. Machine rule had its classic period from the 1860's to about the 1940's. It is now much diminished, though not extinct and at least one machine of classic style survives—Mayor Daley's in Chicago.

A machine is a party organization, but one based on small electoral divisions (wards and precincts) and the trading of services (jobs, welfare, protection, bribery, contact and communication) in return for votes. It was based historically on an immigrant population and corruption. The immigrants needed services and security not political participation. Their needs were met in return for handing over politics to professionals who were interested in power and wealth. The machine was a closed political system, not concerned with issues, debate and agitation or the public interest. But it may well have provided the urban working classes with an efficient system of social welfare and political communication.

The reform tradition arose in part as a reaction to the corruption and inefficiency of the late nineteenth-century machine, but is linked historically with Progressivism and its raw country cousin, Populism. Machine politics developed naturally in the fast growing and teeming cities of the US; its philosophy was invented later. The reform tradition had from its beginnings a philosophy—or several—and publicists to promote it. 'Reform is what America has instead of ideology.' Reformers campaigned for good government recoiling from government by 'a crowd of illiterate peasants, freshly raked in from Irish bogs, or Bohemian mines or Italian Robber nests' (Andrew White in the *Forum*, Dec., 1890, quoted in Banfield Reader, 1961, 213). By good government they meant a reformed political structure—non-partisan and at-large elections, city managers, secret ballot, merit appointments; and progressive aims—the elimination of corruption, efficient administration, welfare policies, civic pride. The public interest was set against the selfish pursuit of private gain. The movement had some affinity with the contemporaneous British gas-and-water socialism (by any other name, of course), the same affection for both welfare and civic pomp (parks with statues, libraries with bell-towers, gothic town halls).

The Reform tradition had its weaknesses; a facile idealism, seeing easy solutions by getting rid of bad men and setting up good machinery; and on the welfare side, a certain chilly Benthamism, again with comparable British examples. Politically, it has lacked the staying power of the machine, but then, unlike the machine, it had objectives beyond mere staying power.

But in the long run of American local government reformism as an ethos rather than a specific programme cannot be discounted. Recent studies have shown the significance of political ethos in determining the performance of urban political systems, particularly in comparison with socio-economic factors. The difficulty here is to isolate 'ethos' from the institutions which it promoted and sustains. But it seems safe to conclude that the two together, reform ethos and reformed institutions, are of potentially high significance.

City government is necessarily a part of the federal system, ultimately subordinate to the State constitutionally, and increasingly dependent on, or at least enjoying, federal financial aid. The autonomy of city politics is further diminished by the incursions of national politics and politicians. America's most successful Reform Mayor, La Guardia, came to power on a wave of New Deal enthusiasm, and was grateful for F. D. Roosevelt's political favour as well as for a generous allocation of federal funds. Federal patronage has been used quite ruthlessly, for example by John F. Kennedy, to influence local parties. Stevenson, a State Governor and Democratic candidate for President in 1952 and 1956, inspired an urban reform wave: a national sentiment gave rise to local action. But national voting sentiment is not transferred consistently to local politics; in particular regular Democrats tend to vote independently for the chief local offices. The reverse impact—of city politics on national politics—has been in the past less dramatic. Mayors have reached the US Senate and one former Mayor of Minneapolis, Hubert Humphrey (a reform Mayor) became Vice-President. Mayor Lindsay of New York is a leading figure in the Republican party but is unlikely to defeat the sitting President for the Presidential nomination in 1972.

There are no academic tests of abrasiveness or demureness. But American city politics—especially big city politics—strikes an English observer as notably less demure than British local government. The stakes are higher for the American politician and citizen: there are greater power and more rewards available. The tensions of low-income whites and Negroes are most severe; and mainly for this reason local politics are often stormier than national politics.

2

The structure of city politics

Areas and administrative units

In a system governed by both the precepts and the aura of the Constitution, the city has no constitutional standing. It is the creature of the states, whose power in domestic matters is still superior to that of the national government. The position is laboriously affirmed in many state constitutions. For example, an article of the New York State Constitution lays on the legislature the duty of restricting the taxing, assessing and borrowing powers of the cities. and adds for good measure, 'Nothing in this article shall be construed to prevent the Legislature from further restrictting the power herein specified . . .'

Judicial interpretation has generally favoured States' rights over the cities. Just as the Tenth Amendment to the Constitution has made the States the residuary legatees of power in the national system, so Dillon's Rule has entrenched the States in the local system. Judge Dillon laid down in 1868: 'It is a general and undisputed proposition of law that a municipal corporation possesses and can exercise the following powers, and no others: First, those granted in express words; second, those necessarily or fairly implied in or incident to the powers expressly granted; third, those essential to the declared objects and purposes of the corporation—not simply convenient, but indispensable. Any fair, reasonable, substantial doubt concerning the existence of the power is resolved by the courts against the corporation, and the power is denied.'

The supremacy of the States has been modified by the pressures of demand and resources (see chapter 3). But the movement for municipal 'Home Rule' has had only modest success. Over twenty states have adopted Home Rule provisions, yet the ultimate bondage of city to State remains, especially in financial matters. Thus, in nominally Home Rule New York State, New York City is still ruled, as G. W. Plunkitt said in 1905, 'by hayseed legislators from Albany', and cannot raise the price of a dog-licence without the approval of the State Legislature.

However, this is not to imply that state governments are highly centralized and coherent bodies, exercising massive power. American local government is, in fact, characterized by a multiplicity of administrative units, over 90,000 compared with 1,400 in England and Wales. The most important are the Municipalities, Counties and School and Special Districts. Most cities (municipalities) are included in Counties, and themselves include School Districts and, in some cases, Special Districts, e.g. for water and sewage disposal. Counties are mainly agents of the State, and operate in an attenuated form within some large cities (e.g. New York, Philadelphia). In other cases (e.g. Baltimore) the city is outside county government altogether.

Most cities have legal boundaries which do not correspond with social and economic activities and needs. Wordy battles have been fought over the proper political and constitutional status of these metropolitan areas, but only three major metropolitan consolidations have taken place (Dade County, Nashville and Jacksonville) and none of these represents such a fundamental restructuring as the Greater London Council or Metro Toronto.

The weaknesses of the political fragmentation of the metropolis are familiar to students of the current proposals for the regional or sub-regional reorganization of British local government. The proliferation of jurisdictions (over 1,400 in the New York Metropolitan Area) gives rise to duplication, overlap and inefficiency; and political energies may be lost through the failure to exploit meaningful communities of interest. Some services are defective because they lack a high population base. In particular, physical planning may be frustrated by the absence of power over the whole metropolitan area (see Wood, 1961, 110-13). Inequities arise from the separation of services and tax-base, especially between the relatively affluent suburbs and the city-centre.

This gloomy diagnosis has not gone unchallenged. Local government has not so far broken down in the metropolitan areas. The requirement of a high minimum population base for some services is disputed (Duncan, 1957 and Dahl, 1967). The tax-base of the city-centre has shown some unexpected signs of buoyancy, and certainly services have been maintained at a level satisfactory to most voters. Fragmentation has been renamed diversity or pluralism.

Such scepticism has been allied to a fresh appreciation of the possibilities of reform short of thorough-going metropolitan consolidation. These include the creation of special districts for particular functions e.g. schools, water, sewage; co-operation more or less formalized by a compact and sometimes based on a federal Council of Governments; annexation by the city of adjacent areas; enlargement of the County's functions within the city; or informal communication and co-operation through officials, parties and the like. State and Federal governments may also exercise a co-ordinating function. However, the variety of devices available is not matched by the energy and frequency of their utilization. There are several reputedly successful schemes of voluntary co-operation, e.g. the Association of Bay Area Governments, ABAG for short, in California. But the most frequently employed device is the least radical—annexation by the city of small adjacent unincorporated areas. None of these devices approaches the creation of government for a metropolitan area, with powers over planning and resource allocation, and involving elective officials and political responsibility: hence the prominence accorded to the few successful attempts at metropolitan consolidation.

In Dade County, the city of Miami and its adjacent county were consolidated as a metro government with powers over water, sewage, transport, traffic, slum clearance and central planning; taxing power was retained by the State. The powers of the Nashville consolidation (Nashville and Davidson County) are not quite so wide. The Nashville plan was passed only at the second attempt. Similar plans have been defeated in St Louis and Cleveland. The success of the reform group in Miami seems to have been due to the absence of any other organized political forces, parties, interest or ethnic groups. Nashville is a more difficult case, with success following failure after the Mayor and one leading newspaper had gone over to opposition. It would at least appear that the support of the Mayor is in some circumstances not essential.

However, the overall failure, not the rare success, of metropolitan reform is plainly the more significant feature. The idea of metropolitan reform has failed because it has little appeal to voters. They do not see the relevance of structural reform to their own neighbourhoods. 'The minute I get out of this city I don't think about it', said one New York commuter. There is evidence that middle-class people of high income and education may look on metropolitan reform more favourably (Booth, 1963, 44-7) and the propaganda campaigns have had some effects on exposed groups. But the majority of citizens are likely to remain uncomprehending and hostile. They have the last word since reform depends on Charter (constitution) revisions subject to referendum, and the best the reformers may hope for is a low poll (in Dade County three-quarters of the electorate did not vote).

Metropolitan reform has failed to catch on to the old reform drives of business efficiency, boosterism and the banishing of corrupt politicians. It runs directly counter to the powerful myth of grassroots democracy. And it fails to face up to the political problems of metropolitan politics: how to solve the genuine conflicts of interest between city centre and suburb and how to avoid a permanent white Republican middle-class dominance. The nation which invented the gerrymander is unlikely to accept readily the imposition of new metropolitan boundaries.

Thus, the city's problems in seeking the power to govern are not confined to its relations with the state. Both within and outside its formal boundaries, the city must concert action with a number of partly independent jurisdictions which have at least the power to delay and obstruct. As with the national political system, the locus of power apparent in state and local government is partly an illusion. Characteristically the system diffuses and attenuates power.

Forms and finances of city government

This softness at the centre prevails too in the political forms of city government. There are four major forms: strong and weak mayor-council; council-manager; and commission. In the usual commission form, executive and legislative power is vested in five commissioners, who are heads of city departments but act jointly as the policy-making body. It is a reformer's system, introduced in 1900 after a civic disaster, and now operating in only 8% of all cities over 5,000. This includes four cities of more than 250,000 population.

The Council-Manager system was also a reform system; introduced in 1908 and spreading rapidly, it now operates in over 1,200 cities, 40% of those over 5,000. A small Council of five to nine members is elected on a non-partisan, at-large ballot (thus happily defeating the machine and its ward-heelers). The Council appoints a professional City Manager, who is, in theory, concerned with administration not policy. In smaller cities and in earlier times, the Manager was often an Engineer or Surveyor. Now, and especially in large cities, he is a professional administrator whose scope extends into the development of policy.

Slightly over half the cities over 5,000 use a form of Mayor-Council government, and this plan, usually with a strong Mayor, is favoured by most of the very large cities. In this form the Council may be larger, up to 40 or 50 in the populous cities; but the median numbers of Councilmen is still well below the British, 13 at highest. The strength or weakness of the Mayor depends on his power to appoint and remove other officials and to supervise administration; his budgetary power and his power to recommend and to veto legislation. Apart from formal authority, the Mayor's power is modified by his relations with party and electors and his position as spokesman and representative, as a focus of the aspirations and responsibilities of the city.

The distinctions of this analysis are a little blurred in practice. There is no clear division between weak and strong Mayor-Council systems; indeed, a particular system may vary with its incumbents and its situation. Some cities are now appointing Deputy-Mayors or City Administrators; thus Mayor-Council systems may acquire City Managers. Since Manager Cities also have Mayors—weak ceremonial figures usually—the differences lie in the distribution of power, not in the presence or absence of Mayor or Manager. However, the true Council-Manager system is distinguished by other characteristics, notably the form of ballot and civil service reform (i.e. merit systems); it has a distinct reform ethos and is conscious of its own political style.

Councils, Mayors and Managers do not comprise the whole of the formal institutions of American City Government. Education may be governed in School Districts, by Boards separately elected and financed. (Only four States and the District of Columbia have no independent school districts at all.) Special Districts have been increasing rapidly since the 1940's and are particularly concerned with fire protection, soil conservation, drainage, water supply, cemeteries and housing and urban renewal. In addition to these independent jurisdictions, city government may itself include boards and commissions with a high degree of autonomy including directly elected or irremoveable heads, access to budgetary provision and federal funds, and their own informal constituencies.

The main source of city revenue is the property tax, which makes up about 40% of revenue; other taxes, including sales taxes, licences and permits, contribute under 20% (Municipal Year Book, 1968, 249). Another 20-25% comes from state and federal grants. The larger cities (above 500,000) tend to raise more per head in taxes, and receive more in intergovernmental grants (Municipal Year Book, 1967, 211). In some cities, but more especially in suburban areas, the property tax is used very sharply, enough to set an English Ratepayers Association on the march. But the great money maker, income tax, is largely preempted by the Federal Government and, on a more modest scale, by the States. City income taxes tend to be too small to solve financial problems, but irritating enough to raise political anxieties.

The major cities do not have the financial resources to solve their problems on their own. They need the States; even more, they need the Federal Government. It is this brute financial fact which now dominates the place of cities in the American federal system, and to some extent the shape of city political systems (see chapter 3).

Even if cities lack money and power, they do not lack functions, some of them carried out on behalf of the State or the National Government. Their most expensive functions are the operation of utilities, especially rapid transit; education, highways, police, fire, sanitation, health and hospitals. Slightly less, on average, is spent on public welfare, housing, parks and recreation.

American cities thus have a range of functions broadly similar to British cities; but there are some differences. American cities have health and welfare functions which in Britain are now largely the concern of the national government. Their police and judicial functions are not separated and centrally supervised, but form a part, in some cities an important part, of local political activity, open to political, partisan and private pressures, and available as a source of rewards and policy goals. This arena of city politics has given rise inevitably to corruption, and has brought saloons, brothels and traffic tickets on to the agenda of politics.

The other striking difference is the absence of the centrally provided and supervised uniformity that is to be found in Britain. American cities are free to be backward if they so choose. Thus, in 1965-6, Boston raised about twice as much in taxation as St Louis, which is about the same size, and spent almost three times as much (Municipal Year Book, 1968, 252-3). The severity of Boston's property tax is notorious. Such differences reflect capacity, needs and demands; the role of the state; and also the ethos, aspirations or style of cities. Some are positive and creative and some conservative.

Electoral systems and parties

Elections are a dominant feature of the American political myth, and there is a wide range of electoral forms and experimentation. Proportional representation by the Hare system of single transferable vote has been used in a few cities, notably New York (1936-45) and Cincinnati (1925-57). Its repeal in the latter city marked the end of its modest vogue. It was advocated by reformers along with at-large elections, the short ballot and city managers, as a way of defeating corrupt party regimes based on ward organization. In practice it seems to have facilitated the representation of minority groups, including Negroes, which led to its abandonment in Cincinnati, and Labour and Communists—which helped to bring about its downfall in New York. Advocates of PR believe it helped to recruit better candidates (and certainly more of them); and to promote issues and party platforms (Straetz, 1958, 268). Its critics deplored the emphasis on ethnic and religious character, the articulation of bigotry. One critic said that candidates were chosen by voters in the order: 'race, geography, religion, fraternal organization, and any specific selfish interest that may appeal to him' (p. 103-4). The evidence of Cincinnati politics in the PR years does not suggest a very high-minded politics; nepotism, corruption, the machine, Communism and the Korean War were all staple issues. But at least the Charter Committee which promoted PR survived to form an effective competing party: thus the city fortuitously gained a two-party system through a device intended to destroy parties.

The Referendum is a normal part of American electoral politics. The voter may be called on to pronounce on state and local constitutional reform including new and revised City Charters; on major financial proposals, especially bond issues; and on propositions, for example, to establish a local library, or undertake a scheme of urban renewal, or regulate the sale of alcoholic drinks, or even to disapprove of US involvement in Vietnam (this one of no practical significance). The referendum endows the voter with a veto power of great significance, which overlays and distorts other formal and informal patterns of power, and affects the outputs of government.

Conclusions on the actual effects of the referendum vary. It brings out the pressure groups, and activates the media, but may fail to disturb public apathy. Many issues are too complex for a voter's judgement by yes or no. There is a tendency to conservatism, and hostility to the paying of taxes, but confusion too, so that a voter may approve a community library, but not the funds to build it. Williams and Adrian found different responses in their four cities. In one city the leadership generally got its way, in another the voters regularly favoured inaction. Willingness to pay taxes rose with social class.

A third electoral device is the non-partisan ballot. This is normally coupled with at-large elections and the short ballot, and has been important in reform theory. Over two-thirds of cities over 50,000 use a non-partisan ballot, and in council-manager systems the proportion is much higher. Strictly, non-partisanship means that there is no party label indicated on the ballot form. Britain has always been non-partisan in this sense, without knowing it, but abandoned the system in 1969. Of course, the reformers hoped to destroy partisan attachments and party organization, but, as British experience shows, everything can carry the party label except the ballot-form itself. The justification for the emphasis laid upon this device lies in the central tenets of the reform ethic. City government is a matter of good business practice and professional administration; parties fall prey to spoils and corruption, and other alleged evils of the machine. If they are concerned with issues at all, these will tend to be national issues, not relevant locally. Instead voters should make individual judgements; candidates may come forward undeterred by the need for party affiliation; and all can unite in the service of the city they love (Lee, 1960, 172 ff).

This picture of non-partisan purity and efficiency is of course less than the truth, though the aspiration itself has a force which must not be discounted. In practice non-partisanship may not make much difference (see Salisbury and Black, 1963). If it does then it tends to create a political vacuum, a lack of structure and organization, which yields to available pressures and facilitates the political influence of a variety of groups, not all high-minded reformers. Non-partisanship gives greater scope to the administrators, and makes easier the taking of positive but unpopular decisions like fluoridation. At the same time it favours the influence of business and the higher social classes, and may thus admit the middle-class values of economic growth and the positive provision of amenities. It permits political action by substantial minority groups like Negroes and Labour. But overall it tends to diminish the electoral expression of social cleavages. Thus far the reformers might be satisfied; but there is a negative side to non-partisanship. It permits the influence of national political affiliation, particularly for the Republicans, whose supporters are more strongly party conscious (cf. Conservative voters in Britain). Liberals in Detroit and Los Angeles have regarded non-partisanship as a way of blocking the Democrats. It promotes a politics of acquaintance and personality, of ad hoc groups and temporary formal organization—and of quite petty issues. It helps the incumbent to stay in office simply because he is known and has experience. It frustrates mass protest voting, since the voter cannot recognize the other side, and it may leave a minority isolated. Indeed some urban experience of non-partisanship demonstrates the force and validity of the usual arguments for competitive party politics: parties do in fact articulate cleavages and develop issues, and provide a measure of responsibility. Non-partisanship, in the light of some American experience, is an attempt to depoliticize what is and ought to be political.

In practice, however, American cities are less non-partisan than the formal count of ballot types suggest. In Chicago, for example, the Democratic machine has triumphed over years of formal non-partisanship; and elsewhere organization and the sheer insistence of issues like race and poverty have repoliticized the non-partisan form.

Partisanship does not always mean competitive two-party politics. Domination of cities by single parties is common—e.g. the Democrats in Boston, the Republicans in Philadelphia—and where patronage remains important the party in office has an advantage to buttress its retention of power. Many of the criticisms made of non-partisanship apply to single party dominance; further, single party systems tend to corruption and lack the reformist drives of genuine non-partisan cities. However, the existence of a vigorous two-party competition at federal, and sometimes at state level, helps to preserve the memory of competition, and encourages the occasional resurgence of the out party.

Innovation is most likely to come from a minority second or third party, fighting for a shift in voting patterns sufficient to give it power, and frightening the incumbent party into change or driving it from office. This role has been played by the Republicans in New York and the Democrats in Philadelphia and New Haven. Beyond this, reform movements tend to produce occasional insurgence, of limited duration. When reform is operationally linked with a coherent political organization as under Charter Committee in Cincinnati or with a Fusion movement as in New York under La Guardia, then it constitutes an effective third party capable of taking and holding power. Generally in the USA there is more scope for insurgent politics than in Britain. There is a tradition of third party movements; primary elections provide opportunity for the insurgent politicians, and primary upsets are common.

Participation in city politics: Councilmen and amateurs

For most Councilmen, as in Britain, local politics is a part-time, unpaid activity; and the Councilmen must aspire to elected executive office in order to acquire a modest salary. In the smaller cities, even the highest executive officer, the Mayor, may be paid a sum which is clearly not intended to be a salary for full-time work. So only a few (though these no doubt an important few in the larger cities) work in politics as professionals, to make a living or to make a fortune. But the financial rewards are now small compared with the great days of the 1860's and 70's, when a Councilman could expect to share in the fabulous proceeds of land and utility franchise sales and one Philadelphia official made $200,000 in a year as the Collector of Delinquent Taxes (Glaab and Brown, 1967, 207).

The political motivations of Councilmen are in consequence of some importance. Apart from the limited opportunities of actually making money, a seat on the Council provides some rewards. For the politically ambitious it is a step to office in the city or to a place in the State legislature and the State party; and thence perhaps to Washington. For the businessman or professional man, including the many lawyers who are Councilmen, the Council offers indirect benefits, prestige, contacts, know-how; or direct opportunities to protect or promote one's interest. For the idealist, there is scope for proselytization and the promotion of schemes of reform.

For such intangible rewards the competition is not always great. There is little systematic knowledge of the processes of recruitment, which appears to come about in several ways. In some cities the dominant socio-economic stratum operates as a recruiting agency, ensuring that its representatives come forward when required. But organized civic groups may also encourage candidature; and formally unattached individuals present themselves, unconscious of the damage they are doing to group theories of urban politics (Polsby, 1963, 31; Lee, 1960, 71-3).

There is indeed, an expectation that politics shall be the concern of interest groups and professionals. There is a tradition of 'joining' but not of voluntary community service in the British sense. Charity was long ago handed over to professionals. The 'service clubs' (Rotary, Kiwanis, etc.) flourish but primarily as 'instruments to enable their members to enjoy the sensations of government while incurring few of its responsibilities'—'a perfect device for political behaviour that is at the same time outside of real politics' (Reichley, 1959, 125). The citizen participant in politics is labelled an amateur, a designation which seems to imply abnormality and incompetence. Williams and Adrian conclude in their study of four cities that rule by amateurs means somebody else must take the decisions (Williams and Adrian, 1963).

Amateurs are most continuously prominent in the

suburbs where 'Residents look to "wheels" to spark civic affairs—men and women who engage in politics as their avocation, and occasionally as their recreation' (Wood, 1958, 175). But amateurs do not dominate suburban politics. In the long run they give way to the bureaucrat, for lack of time and energy and because serious political organization offends the non-partisan, non-political majority. In one field, however, a host of amateurs participates vigorously and sometimes even ferociously—in the very local politics of schools. Here 'grass roots democracy attains its ultimate promise' and citizens 'lay bare their most fundamental convictions' (p. 194).

There is from time to time a resurgence of more widespread and general amateur activity, especially in the Democratic political clubs which have gained substantial membership in some areas. California, for example, had 400 clubs by 1960, with 26,000 members. The amateur of the political clubs is usually a well-educated professional, young enough to be still mobile and childless. Many are women. The style is cosmopolitan and intellectual, though there are more lawyers than academics, except in California, where 'some academics are engaged in politics as a kind of substitute for scholarly research and publication'! (Wilson, 1962, 13). The club-amateur is fascinated by politics, but pursues it also for genuine liberal and reformist aims. Movements of this kind are usually triggered by some local scandal and are stimulated by national political moods: for Democrats in the 1950's Adlai Stevenson was the patron saint.

It follows that the club movement is a temporary fluctuation, bound by time and generation. Amateurs, no longer youngish, mobile and childless, may hand on the torch; but the new generation has different objectives, reacting to Vietnam, not Korea, and failing to find in McCarthy a patron saint of Stevenson's durability. Even at the height of activity the clubs have not had much practical success, and both Daley in Chicago and Wagner in New York preferred to manage without their support.

Interest groups in city politics

City governments have enough scope to attract the pressures of interest groups. Industrialists want sites, department stores want parking lots, private parking companies do not want cheap and plentiful municipally provided parking; taxi-drivers want considerate policemen; labour unions are concerned about work regulations; existing laundries advocate the restrictive regulation of coin-operated laundries; and so on. There is a multiplicity of groups promoting and protecting comparatively narrow interests of this kind, and ranging from the permanent, highly organized to the ad hoc and temporary. Some work through private access to council and officials; but many campaign for public support and may even contest elections. Businessmen, realtors and insurers are over-represented on city councils. These, being generally small, do not represent a wide cross-section of city and neighbourhood interests groups; but, being generally weak, they are not of high strategic importance for the spokesmen of interests. The system of semi-independent boards and commissions offers further openings to determined pressure. Williams and Adrian found in three of their four cities that the planning commission had been 'colonized' by the informal community leaders. Political parties may also serve as instruments of group pressure, particularly in strong competitive party systems, where one party may be effectively the spokesmen of a single group.

Business interests are the strongest, although there are very few examples now of the 'company town' dominated by one major economic interest—the Mellons in Pittsburgh, General Motors in Detroit, Kodak in Rochester, US Steel in Gary, Indiana. The political strength of business varies with the size and nature of the business, whether it is 'home-owned', the number and kind of its executives, and the political tradition of the city in which it operates. There is rarely a homogeneous and indivisible group. There are conflicts of interest between small and large business, manufacturing and commercial, central and suburban. There are also competing interests of high status and with adequate political resources, especially those whom Dahl has called the Social Notables, persons of high prestige, traditionally accorded respect and influence in the community, but not necessarily acting as spokesmen for economic interests (Dahl, 1961, chapter 6). Businessmen who devote much time to politics tend to acquire perspectives and to engage in situations which modify their business outlook.

However, the appearance of business interests checked, moderated and civilized may be misleading. Business goals are frequently congruent with the aspirations of the community. Politicians prudently avoid challenge to strong business groups. The exercise of influence by businessmen does not always require the positive application of pressure; the pressure is already there like atmosphere. The business group can therefore confine itself to deliberate pressure in specific instances; and there is evidence of a high rate of success in this limited kind of action. If there were an open challenge, Dahl concludes (p. 84), then the community would defeat business; the triumph of numbers over notability. Thus, the pluralist democrat has the assurance that the worst cannot happen: dominance by business interests can be avoided by conceding substantial influence peaceably.

The labour interest is by contrast politically quite weak. This weakness reflects the organizational weakness of the unions, their lack of agreement on political action and goals; and the hostility of the prevailing ethos to what is regarded as a narrow and selfish interest. While the craft unions have been mainly concerned with immediate industrial gains, the big industrial unions tend to engage in political activity. The AFL-CIO has a Committee on Political Education (COPE), and this exists in most large towns. COPE functions as an organizing agency for the Democratic Party. In Detroit, with the backing of the highly political Automobile Workers, it has provided leadership, a platform and ward-level organization for the Democrats. It has virtually operated as a political party itself, and recruited middle-class, professional people to broaden its support. Despite this, and considerable success in State and federal elections, COPE has had 'relatively little success in local politics' in Detroit (Gray and Greenstone, 1961, 372). It has never elected a Mayor, and never won a secure majority in the Council. Of course, this is only to say that COPE has not taken over the city government: it has nevertheless been influential, and no Mayor dare ignore it. But it is still difficult to distinguish between the effects of the organization and the effects of the existence of a large number of voters who are also workers. Even without union organization, the Detroit workers have considerable political power through the ballot.

In cities with more diverse industrial structures, labour cannot act as a party. In some cases, e.g. Minneapolis, Houston, San Diego, there have been attempts at a coalition of the unions and reform groups, that is a labour-liberal alliance, in American terms. In New York, the garment workers have organized their own party (the Liberal Party), which has occasionally effected an alliance for electoral purposes, as in the campaign for the Mayoralty in 1965. Most unions in New York and, to a limited extent, elsewhere, support the Democrats, thus trading the advantages of a biddable position, for tenuous acceptance within a middle-class dominated party. However, the more common political pattern is for the unions to act on a narrow front, seeking favours independent of party, and often of each other. Thus, the Construction Unions in New York, following a typical union strategy, opposed the decentralization of schools, wishing to preserve their existing happy relations with the Board of Education, but also opposed the diversion of the Board's funds from buildings to text-books.

Everywhere the Negroes lack influence proportionate to their numbers. Their participation in politics is low, and their share of elective offices is well below their due (Banfield and Wilson, 1963, 293). Only three sizeable cities have Negro Mayors. In terms of policy outputs, the Negroes have made only modest gains in school desegregation, fair employment practices and civil rights.

The political failure of the Negroes contrasts with the success of the immigrants from the mid-nineteenth century onwards in securing a footing in the political system. The explanation lies no doubt largely in the effects of colour prejudice, but is more complex than this alone. First, the machine was already under attack by the turn of the century; and it was by then in the hands of other immigrant groups, Irish and Italians. When a Harlem leader, J. Raymond Jones, finally in 1964, succeeded Carmine de Sapio as leader of Tammany Hall, Tammany no longer counted for much. Second, Negroes do not sustain coherent communities. They lack family structure, and a commitment to neighbourhood and civic activity; they are divided among themselves and separated from other minority groups; there is talk of solidarity but in practice little group effectiveness.

For the most part, therefore, Negro political activity has been limited, sporadic, ill-organized; and, short of militancy, it has necessarily flowed along channels cut and guarded by white men. Some favoured political styles and strategies—Dawson's Negro machine in Chicago and Powell's demagoguery in Harlem—have not achieved much. However, following the Civil Rights legislation of 1957-65, a gradual increase in voting has brought rewards in amenities likes street paving and lighting, and fairer policemen and courts. Specific protest activity, sit-ins and boycotts, have brought some concessions, notably in school desegregation. In some cities NAACP and CORE have been effective. In Detroit alliances with the Unions, in Atlanta with the middle classes, have worked. A decisive factor everywhere has been the attitude of powerful white groups, including Jews. The future of the Negroes in the city is unpredictable, but it is not within the normal tradition of city politics for the liberal conscience to triumph for long over the forces of law, order and inertia.

The pressure of the Negro population is a crucial test for the pluralist ethos, which permits, indeed encourages, the open struggle of groups for influence, expecting that in free competition, a just balance will emerge. There is, it is true, a multiplicity of competing interests which moderate the dominance of the few. The political resources of the business group, though superior to those of any other group, still normally fall short of a capacity to deliver votes or dictate policy. The city government has political resources of its own to deflect if not to repel the incursions of the Economic Notables.

Still, there is no reason to believe that a natural balance of forces arises often, or other than fortuitously. The activities of pressure groups in fact make more difficult the adjustment of the urban political system to social needs, and tend to operate a veto on innovation, driving the city into a disabling immobilism. 'Strong minorities block policy change even more often in municipalities than in the US Senate' (Campbell and Burkhead, 1968, 609).

The press

Compared with Britain there is a vigorous local press in the USA, which is of some political significance. In 1962 there were 1,760 daily papers in 1,400 cities; about 60% of the US population is served by at least two daily papers (Banfield and Wilson, 1963, 313). In contrast the British public is offered 142 national and local papers (1968). The American daily press is declining slowly, but it still operates on a widespread and competitive front.

The effect of broadcasting on the press is still largely obscure. The average American citizen is likely to spend far longer viewing television than reading the paper—according to one survey, a mere 17·4 minutes reading to 2 hours 52 minutes viewing (Banfield Reader, 1961, 385-6). His political information and opinions are unlikely to be much changed by such cursory reading, though the press deals more continuously than radio or television with political news, and some newspapers have a strong political tradition and close links with the community.

Whatever their actual influence, newspapers certainly participate in politics, and there are countless examples of newspaper activity on both winning and losing sides. Thus, the Chicago *Tribune*, consistently Republican in a Democratic city, is often on the losing side; but many city issues do not divide in simple partly allegiances, and even Mayor Daley receives, and will work for, an occasional favourable mention. The New York *World-Telegram* and *Herald-Tribune* both supported the reform movement which triumphed with La Guardia; and the *Times* opposed PR, which was abandoned. In Nashville, both newspapers were active in the campaigns for metropolitan consolidation; but the reform was voted down when both newspapers supported it, and approved four years later, when one paper opposed it. In complex political situations, with success and failure difficult to assign or even discern, the measurement of the political influence of the press is virtually impossible. However, it is regarded as influential by citizens as well as editors, and may be ranked as high in influence as 'Businessmen' and 'service clubs' (Lee, 1960, 77-9).

Press coverage and treatment plainly have some effect on electoral activity. Without newspapers there could be no campaign and no illusion even of electoral debate and contest. Jim Clark, Mayor of Philadelphia and subsequently a Senator for Pennsylvania, said, 'When you got newspaper coverage your campaign is alive' (quoted in Reichley, 1959, 62). In Philadelphia the opposite was a plain possibility: the *Inquirer* 'blacked out' Harold Stassen's gubernatorial campaign in 1958. In some cities success in elections has been firmly attributed to newspaper support, for example, Mayors Murphy, Cobo and Miriani in Detroit. Democratic successes in the 1930's in the face of a hostile press, were still affected by that hostility (Gosnell, 1968, 180-1). Where ballots are long and complicated, a newspaper may be a necessary guide to the confused voter. However, at election times, politics is exciting enough to be dealt with by television, and some victorious campaigns have exploited television, e.g. Collins in Boston (Levin, 1960, 16-23).

Newspapers may have some effect on the tone and content of political discourse. For example, in Cincinnati under PR, the press was accused by reformers of introducing irrelevant national issues, Truman, Korea, Communism (Straetz, 1958). In some cities urban redevelopment issues have been much agitated by the press, which likes to stand for civic improvement, but is limited by the patience of its readers in following what often became highly complex, technical and wearisome stories. Lacking much open and dramatic controversy, reporters have sometimes simply reflected the current mythology about slums and rebuilding, writing from the soft disinterested point of view of a casual by-stander; or, baffled by technical questions, have passed on the Agency's handouts or the hostile propaganda of business interests (Gans, 1962, 172 and 187; Kaplan, 1963, 33).

These examples suggest the nature and conditions, if not the extent, of the political influence of the press. The political activity of newspapers is limited by their business goals; by the journalistic problems of reporting and editing, and by the difficulties of communicating complex matter to large numbers of readers. A newspaper is rarely an entirely free and efficient instrument of its proprietor's or its editor's views. In this situation newspapers tend to neglect politics altogether, or to dress up politics as a crusade for civic virtue and pride against corruption and inefficiency. By stressing the wickedness of politicians and the critical nature of the times, newspapers may in fact deter the citizen from an interest in politics, driving him to abstention and hostility.

The extent of the influence of the press varies according to the political culture and structure of the city, and the issues confronting it. The ubiquity of elections and the tradition of public political campaigns give scope to the press; so too do public hearings and referenda. Where politics is highly organized on a partisan basis, the press has minimal influence. In weakly structured and nonpartisan settings, voters may look to the press for cues. Where politics is concerned with the distribution of favours to influential groups, the press claims a share of favours. Where issues are visible and open to press exploitation, then Mayors are vulnerable.

However, these conditions need to be interpreted in the context of the complicated processes of communication in city communities. Scott Greer concludes (1963, 190) that 'a political process does go on at the grass roots', but also (p. 193) 'at the grass roots, the voters did not know what was going on'. The citizen reads (a little), talks and reflects (but not much about politics). The press both helps and hinders him. Politicians worry about what the papers say. They are probably right to do so; but perhaps they ought not to read the papers, sport and all, for more than 17·4 minutes each day.

3

The city in the federal system

The nature and development of the federal system in relation to the cities

The American federal system cannot be defined in terms of hierarchies or autonomous political units. It is a sharing in government by diverse territories. The citizen belongs to at least three such territories (Federal, State and local) and possibly more, e.g. county, school or special district. His taxes go mainly (64%) to the Federal government, but are spent in part by the State and locality. The matters which affect him most, defence and the state of the economy, are federal. But those which affect him most directly are State and local. Responsibility and influence flow in all directions, there is overlap as well as interlock, and mutual frustration as well as reinforcement. Conflict is rather more characteristic than collaboration, but the system does not actually fall apart.

Historically, the relative parts in the system of Federal, State and local governments have fluctuated. The Constitution is an invitation to a struggle, from which the Federal government has gained most. Economic and international circumstances—wars and welfare—have forced a centralization of policy-making. Dual federalism (which meant, 'leave it to the States' or 'leave it alone') prospered from 1860 to 1937 but collapsed at the height of the New Deal, with a shift in the majority judgements of the Supreme Court. The 'co-operative federalism' which succeeded it recognized the legitimacy, even the desirability, of federal involvement in domestic affairs. 'Creative federalism', a fashionable phrase ascribed to Governor Rockefeller of New York in 1962, but popularized by President Johnson, marked a new, possibly temporary, enthusiasm for positive government. But this is not simply a further stage in the inexorable enlargement of the power of the national government. States can be creative, too, so they say, and the federal system is not after all, on the point of withering away.

The States have never lacked authority and honour in the system. They began as independent units; constitutionally they received substantial reserved powers and a role in the amending process; some grew to be as large as modern nation-states (e.g. California and New York are almost as populous as Canada, Illinois as Belgium). Despite the twentieth century ravages of wars, depression, income tax and New Deal (Fair Deal, New Frontier, Great Society), their role in domestic expenditure has remained a prime one. State-local expenditure for civil functions has normally been higher than that of the National Government and this is still the case.

The States have had a legitimate and traditionally respected position from which to revive their fortunes. The cities, by contrast are parvenus, coming up with perhaps unjustified self-confidence from origins which constitutionally were humble enough, though socially and politically of high significance. The dominance of the cities by the States was asserted in the famous judgement by Judge Dillon in 1868. 'Dillon's rule' has been challenged, on political and legal rather than constitutional grounds—

the inherent right of local self-government, the common law rights of corporations—but has never been overthrown. Many State constitutions bind subordinate local jurisdictions especially in financial matters; and for many cities 'Home Rule' is still a battle-cry rather than an achieved status.

Nevertheless, social developments have undermined constitutional certainties, as they usually do. Poor people live in crowded cities and their problems and their votes have modified federal relationships. Governments, national, state and local, faced with urgent social problems, have instituted massive programmes of expenditure for social welfare. These have been operated and funded either directly from all three centres, or by means of intergovernmental payments, nation to state, state to locality, and nation direct to locality. The amount of Federal aid to state and local governments has jumped since the mid-fifties: from $3·7 billion in 1956 to $12·9 billion in 1966, and with an estimated further sharp rise to $17·4 billions in 1968. The scale of total expenditure and of financial aid, and their rate of increase, might have destroyed the balance of the federal system. This has not been the case, however, for since the 1950's State and local expenditures have also risen roughly in proportion with each other and with the federal share. Of course, influence in the system is not exactly proportionate to dollars disbursed. There seems to be a threshold beyond which relative influence depends on political as well as financial factors. ('The man who pays the piper . . .' is much too simple a notion in public administration.) Still, the stability of financial relationships has tended to stabilize the whole system. Federalism has been modified but not transformed.

Table showing the percentage distribution of direct spending for general domestic programmes (excluding defence, space and international programmes; also trust funds and government-operated enterprises).

Fiscal year	*Federal*	*State*	*Local*	*Total*
1965	34	23	43	100
1960	36	22	42	100
1955	38	21	41	100
1950	46	19	35	100
1944	60	12	28	100
1936	49	15	36	100
1902	28	9	62	100

Source: Tabulations of the Governments Division, Bureau of the Census, printed in Revenue Sharing, 1967, 160.

Table to illustrate growth rate of civilian-domestic direct general expenditures by governments since 1948 (excluding defence, space and international expenditures; also interest on general debt and veterans' services).

in billions of dollars

Government	1948	1954	1964
Federal	8·7	12·8	22·8
State	6·2	10·1	24·3
Local	11·5	20·6	45·0

Derived from Table 1 in Revenue Sharing, 1967, 225.

Working relationships

(i) City and State

Cities and States have been locked in uneasy relationship since the Colonial period. The unease is the product of economic, social and demographic disparities. These are most strikingly seen in the relationship of New York City to New York State—a world city of ten million people subordinate to a State capital in a small town over one hundred miles away—as if London should be ruled from Leicester. The power-relationship is not exactly the same as the geographical but to a frustrated Mayor of a near-bankrupt city trying to get his Budget through the State legislature, it may seem so.

The state legislature governs the City through statutes, appropriations and investigations. The Judiciary provides an authoritative, detailed and persuasive review of municipal activity. The Governor and the State administrative branch operate a range of checks, reviews, vetoes, approvals and orders, supported by some powers of appointment and removal, and of direct administration.

These formal bonds are moderated by personal relations and political processes. Governors and State legislative leaders may establish positive if not cordial relations with Mayors and other City leaders. Shared party loyalties may, but do not always, lead to co-operation for common goals. Indeed, both party and goals may well differ between the differing electorates of city and rural or suburban State. City pressure groups may enlist state support against the Mayor. But the Mayor is not entirely without political resource: his electorate is part of the State's electorate, the City has its delegation to the State legislature, the State needs the City's administrative machine. Still it is not surprising that Mayors, individually and through their organizations, express dissatisfaction with their bondage to the State.

Many of the States have serious political weaknesses: excessively detailed constitutions, weak executives, inadequate fiscal resources and part-time legislatures biased by malapportionment towards rural areas. They have neglected the cities. The former Secretary of the Department of Health, Education and Welfare (HEW), John W. Gardner, described their dealing with the cities as 'inadequate'. In theory, State governments are in a position to be of great help to the cities, but for the most part they have not been so' (Federal Role, 1966, 315). Mayors have spoken more sharply. Jerome Cavanagh of Detroit told the same subcommittee a few days later in answer to a question by Senator Robert Kennedy about what the States could do for the cities: 'I know I would speak for every mayor of America, in saying that States certainly could and should do much more than they are presently doing . . . the States have almost literally turned their backs on the problems of people that live in urban areas' (p. 631).

Of course, Governors have testified on the other side. For example, Governor Laxalt of Nevada wrote to Senator Muskie, Chairman of the Subcommittee on Intergovernmental Relations: 'The presumption that there is in state capitols a singular lack of knowledge, expertise, and maturity with which to meet social and other problems locally is quite erroneous' (quoted in Creative Federalism, 1967, 634). His solution was to return to the States 'without strings—a portion of the enormous Federal revenues States send to Washington'. There is indeed some evidence of State initiative in the reform of its procedures for City government, in providing for consolidation, annexation, equalization for example, and in raising and disbursing taxes which have increased since 1948 about twenty times. By the end of 1966, ten States had established Cabinet level Departments for Urban Affairs, and the number is now much higher. Some States have creditable records: New York is one such but even here the City's case for a wider measure of Home Rule was rejected by the Constitutional Convention of 1967.

The position is not static. The State system is not of its nature incapable of producing viable and effective government. It seems likely that State government is on the edge of radical change; Mayor Collins of Boston, a noted urbanist, arguing for direct access to Federal grants, expressed both the dissatisfaction and the expectation of change to Senator Muskie's Committee on 'Creative Federalism'.

'Despite reapportionment of State legislatures, despite increased recognition by Governors that they are no longer elected by rural minorities, despite increased efficiencies in many States departments and agencies and despite other heartening factors, State government in Massachusetts and most other States is not yet ready to properly assist local governments meet the challenges facing them' (Creative Federalism, 1967, 653).

Working relationships

(ii) City and Federal government

The City's relationship with the Federal government is already deep in a revolution. In 1955 a study report to the Kestnbaum Commission on Intergovernmental Relations stated: 'It is for the states to take an interest in urban problems, in metropolitan government, in city needs. If they do not do this, the cities will find a path to Washington as they did before, and this time it may be permanent . . .' (Kestnbaum Report, 1955, 40).

The path has been found, as much by Federal administrators as by the City leaders. Until the 1930's, the Federal Government was mainly concerned with education, agriculture and highways. In the 1930's a range of welfare and economic security measures was launched, and these have been developed rapidly in the 1950's and 1960's. Total aid to metropolitan and urban areas was about $3·9 billion in 1961 and was estimated to reach over $10 billion in 1968. Most of this comes, either through the State or direct, in the form of a grant in aid, assigned for a specific purpose and with conditions about 'matching' finance and administration.

As with the States, the City-Federal relationship follows the pattern of the three branches of government, but has recently been under greater pressures of growth and adaption. In the New Deal period, the American system of government was radically changed by the demands of positive government. From the 40's to the early 60's the system was transformed again by the demands of foreign and defence policy-making. From the 50's through at least to the 70's, the system appears to be going through a period of further change, brought about by domestic crisis. The emerging picture is of a system under stress.

Several Departments and agencies have programmes with a significant urban impact: notably Agriculture, Commerce, Labour, Transportation, HEW and above all Housing and Urban Development (HUD) which alone accounts for almost half the total aid in the mid 1960's; and among agencies the Office of Economic Opportunity (OEO), the Small Business Administration and the Veteran's Administration (which in 1964 spent over $3 billion on programmes of this kind). Most of this urban aid is channelled through the States to the Cities, but a proportion of it goes direct to the cities. In 1964 about a tenth of the total federal aid to State and local governments was actually granted directly to local governments, and this fraction has been growing. Some major programmes, like that for Model Cities, have been specifically designed for direct Federal-City administration.

There are two striking characteristics of this inflating activity in Washington DC. The first is its ill-co-ordinated multiplicity. Altogether nine departments and seven agencies operate programmes with 'significant metropolitan or urban impact'. In some cases different agents deal with related parts of the same problems, for example, in highway and mass transit facilities; and in income supplements and public housing. Co-ordination has been attempted by HUD (under the authority of an Executive Order to convene meetings with other agencies); by the Bureau of the Budget and OEO; by the regional offices of the Departments; and by the President's arrangements for personal liaison, through the Vice-President or a specially appointed 'ambassador' to the Cities; and now more formally, through the Council on Urban Affairs established in 1969.

However, these efforts have not silenced the complaints of the local governments about lack of co-ordination. An extra complication has been the use, especially by OEO, of organizations separate from local government altogether. The Mayor of Springfield, Tennessee, accused the Office of being 'sometimes suspicious of, or hostile to, local government' (Creative Federalism, 1967, 667).

The second characteristic of the new federal-city relationship is the expansion of the cities' lobbying in Washington. The larger cities now maintain offices in the capital, but Eastern cities have always taken advantage of their proximity to Washington. Mayors go the rounds of the departments, pressing their claims, but baffled by the size and complexity of the government machine. Sometimes they may mistake courtesies for concessions. The Mayor of Omaha reported: '. . . Immediately after taking office, I came to Washington, to Mr Weaver's office, with a plan for alleviating the problems of my minority housing. Mr Weaver was kind enough to say that some of these ideas were unique, but by the time it filtered down to the man who has to implement the idea, it became a bad idea . . . there seems to be an obstacle course between their office and the city of Omaha . . .' (Federal Role, 1966, 1052).

Apart from these individual forays, the Mayors are well organized in the US Conference of Mayors and in the

National League of Cities. It is conceivable that these direct links, mainly between professional administrators, have created a new focus of influence, not organized around levels or branches, but vertically among the bureaucracies.

The legislature has trailed behind the Executive in providing for the cities. Lacking the Executive's capacity for initiation and multiplication, Congress has improvised its reaction to the urban crisis. Legislation has characteristically lacked cohesion, yielding by instalment to the Administration's pressures. The committee system, which is the main instrument of Congress, has not been adapted to the new cluster of urban problems. Thus, the new Department of Housing and Urban Development has working relations with three committees in the Senate; similarly, metropolitan transportation is dealt with by the Committees on Banking and Currency (mass transit legislation), Commerce (commuter railroads) and the Public Works Committee (highways).

Still, failure to adapt a committee system has not invalidated a considerable effort by Congress to confront the problems of the cities. Since the Watts riots of 1965 inaugurated the contemporary urban crisis, the major committees have held series of special hearings, in addition to the regular legislative and appropriations sessions. Senator Ribicoff's Subcommittee on Executive Reorganization produced over 5,000 pages of testimony in hearings in 1966 and 1967. Another Subcommittee of the Senate's Committee on Government Operations, that on Intergovernmental Relations, Chairman Senator Muskie, has investigated 'Creative Federalism' (1967); and the Joint Economic Committee has looked at Revenue Sharing. Hearings of this kind reach a small but important audience and at once expose and modify the context of opinion lying between Washington and the cities.

Committees have considerable influence because of their power over legislation and appropriations. True, in President Johnson's golden years (1964-66), the Administration often had its own way. Senator Ribicoff told his Committee in December 1966: 'One of the great problems that bedevils Congress is the fact that the entire legislative process is now in the hands of the executive branch.' There was pressure, he said, to hear only Administration witnesses, already committed to the Bill, and then to proceed to 'mark up' the Bill and get it 'out on the floor'. However, this is a possible but not a normal relationship, as the legislative frustrations suffered by Johnson in his last two years show.

One recent committee hearing provided a striking though uncharacteristic example of a confrontation between Congress and City, Senator and Mayor. Mayor Samuel Yorty of Los Angeles (which includes Watts) clashed with Senator Robert Kennedy. Here is part of the exchange:

> Senator Kennedy: . . . You cannot talk about the unemployment rate in this area and in a very inflammatory situation and say it is somewhere between 15 and 35 per cent. You might not have the responsibility in each of these fields, but you certainly are Mayor of the city and therefore we need some leadership.
> Mayor Yorty: I do not need a lecture from you on how to run my city. I think you should confine your questions to things that are possible for me to answer without bringing a computer. . . .
> Senator Kennedy: . . . these people expect to have as much of a chance as you and I have had.
> Mayor Yorty: Well. Certainly they will not have the chance you have had, but I hope they have the one that I have had . . . (Federal Role, 1966, 764 and 771).

Confrontations in committee hearings are rarely as heated as this, but sharp exchanges are certainly more frequent than in the comparatively demure Select Committees of Britain, which rarely hear testimony from politicians. The Kennedy-Yorty encounter indicates the possibility of political tension in federal-city relationships, and in this case perhaps, the complication of the relationship by political ambitions connected with State and national party politics.

The impact of the Federal Supreme Court on municipalities follows the lines of its impact on the States. Under the Warren Court, cities have been affected by Court rulings on school desegregation, and on the right to vote, the regulation of wages and hours, and questions of unjust dismissal of employees. The functions of the city in matters like housing, planning and minor commercial activities, and the division of power between municipality and state, and the limits of Home Rule, are adjudicated from time to time. There is no escape from judicial review in a federal system based on an obscure but revered Constitution, and a traditional abhorrence of central power. But the Federal Supreme Court has upheld Federal intervention and regulation for over thirty years; and even with the post-Warren Court more sympathetic to States rights, the precedents and the pressures may not allow more than modest increments to the powers of the States. Such increments may however be won from the municipalities as much as from the Federal Government.

Cities and the Federal Government: the bargain appraised

The Federal Government has provided financial aid for the cities in a time of desperate need. Mayor Lindsay's views are widely held in City Halls:

'. . . With taxes already at a level which causes daily complaints by both business and individuals, we are barely able to meet our essential needs. New taxes, higher taxes, are no solution in New York, and to a greater or lesser degree, I expect that the same is true for our other cities. The money must come from Federal revenues. Without very large amounts of Federal money, the plain fact is that the crisis of our cities will continue and worsen' (Federal Role, 1966, 561).

The Federal Government alone has a sufficient tax base in income and corporation taxes. The States rely mainly on sales and automotive taxes, sometimes supplemented with a small income tax. The inelastic property tax is left to the localities, as in Britain, which may add sales and other taxes, including a miniscule income-tax. As in Britain, local taxation is highly 'visible' and politically sensitive, and cannot be much expanded without major changes in public attitudes.

In providing finance, the Federal Government has also done a little to equalize conditions and opportunities between the States. Their social and economic disparities are substantial. For example, illiteracy ranged from over 6% in Louisiana to less than 1% in Iowa and several mid-Western States (1960). Current expenditure per pupil in public schools ranged from $3-400 in the South Eastern States to $876 in New York (1956-66). The process of equalization is unsystematic, arising from the direction of particular programmes to areas of need. Obviously it is far from effective: it moderates but does not remove gross inequalities. Still the process, from a taxpayer's point of view, may seem radical enough. According to one study, 'For 1965 the people of Illinois, Indiana, and New York, for example, paid about $1·60 in Federal taxes for each dollar of nonhighway aid while for Arkansas, Mississippi and Oklahoma the figure was around 30 cents' (Tax Foundation, 1966, p. 3). This measure of equalization must rank as a considerable achievement in a polity where inequality between States is still regarded as a matter of States Rights, and equalization is not generally accepted as a proper objective of national policy.

An assessment of the actual achievement of federal programmes in the cities is more difficult. In a notorious judgement on urban renewal, Greer wrote: 'At a cost of more than three million dollars the Urban Renewal Agency (URA) has succeeded in materially reducing the supply of low-cost housing in American cities' (Greer, 1965, 3). This was true, though to be fair, the production of low-cost housing was not the sole, or even the primary objective of the programme. But such judgements are premature in the face of massive problems, with solutions possible, if at all, only in the long term. In the case of urban renewal, the programme has been modified and improved in the 1960's, with a greater emphasis on rehabilitation and conservation, on the needs of smaller cities, and on a simultaneous attack on all the social problems of the slums. President Nixon's Administration is at least considering the possibility that the most effective way to alleviate poverty is to give the poor money. However, it is unlikely that federal-local programmes in the major fields of housing, welfare, transit and education will be much diminished.

The Mayor of Atlanta, facing problems of massive immigration, cheerfully supported Mayor Lindsay's gratitude for federal financial aid. 'Federal encroachment', he said, 'is federal salvation.' But, of course, the reverse is true too, and the cities have paid a price for their salvation. They have, first, lost freedom of choice in policy areas where federal aid has been accepted. Few grants are without strings, and block grants are a Mayor's dream not often come true. The reality, according to the Administrator of Boston Redevelopment Authority, lies in: 'Our thirty-odd year old philosophy that Washington knows best, that the cities and states are not to be trusted to use their own discretion (This) has created a vast advice-giving, project-application-reviewing bureaucracy in too many of our federal aid programmes. The attitude . . . is stifling local initiative, keeping young talent away from local government, and stretching out local performance schedules' (Federal Role, 1966, 2796).

A second item on the debit side, at least as seen from City Hall, has been the remoteness and lack of understanding of the Federal Government. Washington, it seems, has not always known best: it has suffered, its critics say, from an 'innocence' of local conditions, from 'tunnel vision'—seeing one problem at a time and out of context. There is little doubt that some federal policies have actually harmed the cities. Federal grants for freeways, rather than for public transport facilities, have failed to solve the city's transit problems, and have intensified the flight to the suburbs. This was already being encouraged by federal guarantees and insurance of mortgages. Meanwhile other programmes tended to concentrate the indigent in the central cities but have failed to secure the rehabilitation of the poor.

The third major defect of federal aid has been its fragmentation. City and State administrators claim with justice that it has been difficult to know about, still more to utilize the 400 Federal programmes already available by 1966 (the total depends on how they are counted). Mayor Lee complained that one attempt at a locally co-ordinated programme for low-income families in New Haven depended on eleven different Federal units and 'it took twenty months to get them to work together with us' (Federal Role, 1966, 912).

Of course, there is an element of unfairness in these criticisms of federal-city relations. The problem of co-ordination has arisen largely from the very energy and inventiveness of Washington. The failure of the Federal Administration to understand all the consequences of its policies has been matched by failures in understanding at local level too. If Washington has not known best, it is not clear that City Hall or State Capitol has known very well. Finally, the loss of freedom has been necessary for accountability, and has had positive effects in the improvement of for example, local personnel and planning policies. Federal salvation of the cities has been bought at a price, but not perhaps an excessive one.

The impact of federal programmes on the urban political system

Federal intervention has introduced an initiating and innovative force into political systems which have otherwise often lacked directing power. The tendency of city politics is to fragment and diffuse power, to accommodate prevailing pressures and to minimise functions. The major federal programmes, e.g. in housing, welfare and urban renewal, encourage, even require, the collection and focusing of political power for positive development. Financial aid is encouraging enough, but there is also the stimulation of output-oriented federal officials, selling nationally promoted policies, and armed with specific federal authority in personnel practices, code enforcement and compulsory land acquisition. Of course, federal intervention is still in most fields virtually by invitation only; but the prospect is enticing, and the cities are inclined to invite the federal government in, and to accept for better or for worse the consequences for their political systems.

Many federal-local programmes are in policy areas where traditional ideas and attitudes are hostile to positive action. Despite the refurbished urban progressivism of the Kennedys, and the fashion for posing the problems of the city (even in the popular weeklies), there is still a good deal of incomprehension, passivity and conservatism. The operation of a federal programme requires the organization of political forces to neutralize or convert the potential opposition. In urban renewal, welfare and the like, 'you can only do what your political structure allows you to do'.

The coalition of political forces is most likely to be centred on the executive power. Some big city Mayors, notably Lee of New Haven, Collins of Boston, Dilworth of Philadelphia, have demonstrated the capacity of their office for combining political and administrative leadership towards positive goals. Mayor Lee is perhaps the most distinguished of the Mayors associated with urban renewal. In New Haven 'very little happened until redevelopment became attached to the political fortunes of an ambitious politician' (Dahl, 1961, 115). It became the central policy of Lee's administration, and after the election of 1955 he concluded that 'the political appeal of redevelopment far exceeded any other conceivable issue within his grasp' (p. 121). Lee was following the strategy of many successful and positive leaders—'picking winners'. (Weaker politicians concentrate on avoiding losers.) Urban renewal looked a potential winner in the old tradition of good and visible government. Fine building appealed to local pride and decency: it was non-partisan, permanent and a visible reminder of its progenitor, like La Guardia's Airport and Moses' bridges.

Most Mayors who have been active in programme politics have been supported by able and influential administrators—Logue in New Haven, Bacon in Philadelphia. In some cities an administrator has been the driving force; a notable case is Louis Danzig in Newark (Kaplan, 1963). The administrator may have the political resources for the successful promotion of a federal programme: a semi-independent base, an invulnerable expertise, established lines of communication to Washington and its Regional Offices, and strategic alliances with other local agencies. The building up of piecemeal commitments within this specialized system has often produced 'the non-amendable package' of federally supported policies.

The executive-centred coalition has had some notable (and well-publicized) success, but is usually subject to challenge and to fluctuation and dilution in power. Some Mayors have suffered electoral reverses, partly explained by their association with federal development; others, more prudent, like Daley, have avoided firm commitments. Administrative power has generally been dispersed in independent agencies serving narrow policy fields. Public enterprise has been frustrated by conservative distrust and by the overwhelming magnitude of the problems. Economic interests have been conciliated at the least, and sometimes accorded dominant influence. Through representation on planning, development and similar bodies, businessmen have been able to promote, modify or veto some federal schemes. In any case, urban renewal and other programmes involving physical plant ('hardware') have been shaped by market forces and largely carried out by private developers. Thus, businessmen have had their triumphs, if not as politicians, then as businessmen.

The federal government's insistence in many of its programmes on citizen participation has failed so far to transform, or even to change much, the local political system. Citizens' committees and community corporations have been established but their impact has been limited. In New Haven, for example, a Citizens' Action Committee was established to represent many of the major centres of influence or status in the community. But it was used by the Mayor to sell his policy; it 'never directly initiated, opposed, vetoed or altered' any major proposal of the Mayor or his Development Administrator (Dahl, 1961, 130-2). Neighbourhood organization was even less effective. Programme promoters are more concerned to build support, and fear with some justification, that geniune citizen participation would delay or frustrate their projects.

Altogether, the record of federal-local programme activity raises doubts about the capacity of the urban political system; but does not suggest that the urban political system necessarily lacks the power to achieve the far-reaching aims of federal programmes. Much of what goes on in a community is not the product of deliberate political decision, and many deliberate decisions have unintended consequences. There are many actors, pursuing multiple and obscure goods, and accommodating themselves not always consciously to the pressures inherent in their situation as they see it. Even the most powerful actors are subject to some extent to the determinism of the game; yet some of them from time to time prove themselves match-winners in their own right, and federal backing has a part in their victories.

The consequences for the federal system

(i) The expansion of federal aid to local government has had important consequences for the federal system. It has opened local government to new national pressures and national government to new local pressures. Urban affairs now occupy a central place on the electoral agenda. In the 1968 Presidential Election, the urban crisis, linked with the riots and the issue of 'law and order' came to dominate the campaign. Electoral pressures conflict: core-city voters demand welfare, while suburban and rural voters want prosperity and low taxes. For all voters, housing and its associated aspects—mortgages, amenity, respectability—matters a good deal. The failure of the Democrats both in the mid-term elections of 1966 and again in 1968 may be interpreted as a verdict, if a narrow one, on the urban programmes of the Johnson administration.

Besides electoral pressures, there are the pressures of the national lobbies, which, even when they represent local forces, tend to shift the locus of decision to their national battlefields. The lobbies include the organizations of States and Governors, Municipalities and Mayors; the numerous professional organizations of administrators, urbanists, social workers, policemen, planning and finance consultants; and the economic interest groups, e.g. Small Property Owners Association, American Property Rights Association, the National Parking Association.

American government is notoriously open to pressure, and there is little doubt that such organizations bend and dent federal and local policies. Thus, a recent study of educational policy has suggested that the Act of 1965 providing for federal aid to elementary and secondary education 'grew out of national rather than state and local political forces. The Act came about when denominational groups moderated their differences sufficiently to co-operate with secular groups pressing for federal aid' (Federal Role, 1966, 2794).

This nationalizing of political pressures on urban affairs has both energized and confused the issues. Daniel P. Moynihan, who has served Democratic Administrations and now serves President Nixon as Assistant on Urban Affairs, has pointed sadly to the conflict between 'political rhetoric' and 'evaluative research'. The difficulty is that the evaluative research rarely provides convincing answers to problems which are partly political, in the sense that they yield to the compromising of interests rather than the more precise formulations of researchers.

(ii) In addition to changes in the political context of federalism, the expansion of federal urban programmes has produced a cluster of problems of management within the system. Solutions have been offered and applied with an energy which does credit to the administrators and the academics involved. The academics in particular, having abandoned prescription in their scholarly research, have lavished it on Washington. (Government by academics armed with evaluative research is surely nearer in the USA than in any other country.)

Co-ordination has become almost an obsession with administrators concerned in federal programmes. Nevertheless, its problems still look too difficult for the proffered solutions. Both HUD and OEO already have co-ordinating responsibilities. President Nixon has established a National Council on Urban Affairs, similar to the National Security Council and located within the President's Executive Office. Other remedies include the consolidation of grants, and a more vigorous attempt to co-ordinate through the regional offices of the Federal Government. Inevitably, in a system so conscious of the boundaries of jurisdictions, there is no enthusiasm for regional co-ordination by federal administrators. There is no agreement about the level or levels at which co-ordination should take place, and an uneasy awareness that the administrative costs of complex operations may be high. Systems of planning-programming-budgeting (ppbs) clarify procedures and objectives, without solving the whole problem.

Personnel problems are prominent because the operation of the programmes is voracious of skilled administrators, both in Washington and locally. It takes a very high quality of city administration to accomplish the work needed to elicit federal aid. In any case 'human resources' figure largely in American administrative theory. If this concern is correct, then there is a serious shortage of skilled manpower, especially at the State and local level.

Fiscal devices provide a way out of problems which perhaps can be evaded but not solved. Tax-sharing, that is returning part of the product of federal taxes direct to States, would enable the States to take over some Federal programmes. John Collins, then Mayor of Boston, described this with some exaggeration as 'the most dangerous idea that has been suggested in America'. He, and many other Mayors, would prefer block grants direct from the Federal Government to the localities, based on an assessment of need and demonstrated performance. The difficulty here is that the central government, as in Britain, does not quite trust local governments with large sums of money. Such distrust may be reinforced by occasional evidence of the use of Federal grants wastefully, or to replace or reduce local taxation.

Another way out, advocated by President Nixon in his election campaign, is to induce private enterprise to tackle the problems of the cities, particularly by providing employment for the poor—'not more millions on welfare rolls but more millions on payrolls'. The emphasis derives in part from the propaganda demands of the election campaign of 1968. It is true, however, that private enterprise has worked effectively on urban renewal schemes in Philadelphia, Boston and Pittsburgh. Senator Javits has proposed a public corporation, the Economic Opportunity Corporation, to stimulate private effort. These and other proposals, including some made by Senator Kennedy, together with the urban operations of Ford and other Foundations, provide an attractive diversification of the Federal programmes; but they do not solve, and may actually complicate, the problems of co-ordination.

These may indeed be insoluble; and, in the end, it may not matter much. The practitioners seem sometimes to be saying:

'For forms of government aid let fools contest,
What e'er is best co-ordinated is best.'

This seems unlikely. The Program's the thing, and there are limits to the blessing of co-ordination.

(iii) A further consequence of the expansion of federal-city relations is to raise questions about the management of the federal system itself. The equilibrium of the system has been disturbed, and adjustments are in train.

The contenders for power in the system do not lack protagonists. The States' case was formally re-asserted in the Report of the Commission on Intergovernmental Relations set up by President Eisenhower for that purpose (Kestnbaum Commission, 1955). 'The National Government need not do everything that it can do . . . Where National action is desirable, greater attention should be given to minimizing its extent and to leaving room for and facilitating co-operative or independent State action.' 'National action should be the least that will ensure results.' Nevertheless, federal-grants-in-aid were specifically approved ('a full matured device of co-operation government'); and the subsequent action proposals for new state responsibilities were limited to programmes of vocational education, and the treatment of sewage.

The case for the National government was meanwhile argued in the footnotes of the Report, mainly by Senators Hubert Humphrey and Wayne Morse. Their case was that the States and local governments on their own would not or could not do enough. On education, for example: 'We believe that the needs of the Nation for an educated citizenry, and the just claims of every child to a fair chance to get an adequate education, outweigh the arguments in favour of complete State and local support for education' (Report, 195n). Senator Humphrey has been able since then, as Senator and as Vice-President 1964-8, to do rather more than write notes of dissent in order to support the role of the Federal Government in the system.

The case of the cities has been fought by the Mayors. Progressive and powerful Mayors like Collins and Lindsay favour direct relations with the Federal Government, seeing the States as 'one additional level of authority and potential obstacle between the place where the problems are and the place where the money is' (Mayor Collins in Creative Federalism, 1967, 658). Federal encroachment, as the Mayor of Atlanta said, is federal salvation.

However, the conflict is likely to be compromised far short of the victory of any one level of government. No side can win outright victory, so it is best that all shall have prizes. The Federal Government (which has the resources to fight the hardest) has rejected the direct relationship with the local governments, wholly excluding the States; still more, the direct rule through its own network of local instrumentalities. Instead it has plumped for a conception of 'creative federalism' defined by Secretary Gardner as the strengthening of State and local government, so that it may enter into a healthy partnership with the Federal Government.

By 1968 theory of 'Creative Federalism' had a serious content as well as a readily saleable label. It includes the notion of a substantial upgrading of State and local governments; active rather than passive government at all levels; the concentration of responsibility in politically responsible and responsive units and officials, in contrast to reliance upon functional authorities and special districts; and for the National Government a supportive and stimulative role in a co-operative system which recognizes that contemporary domestic problems need not, possibly cannot, be solved at one level alone. (See Creative Federalism, 1967, 914-5). It is by no means clear that this theory will survive unchallenged under President Nixon.

The balance of the federal system has been shifted by the impact of the urban crisis, but not destroyed. The system is indestructible because it is entrenched in history; because it reflects an American philosophy of decentralized government and diffused power; because it is supported by the necessary and continuing courtesies of bureaucratic co-operation; and because under pressure it is capable of adjustment. To judge that the federal system is a viable system tending to equilibrium is to say more than the system is not actually falling apart. In that durable framework the cities are firmly fixed, trapped or cradled according to viewpoint.

4

Four political roles

(1) *The Mayor*

The city Mayor is the most visible and newsworthy actor in American local government; contrasting by his political involvement and apparent executive responsibilities with his ceremonial British counterpart. In fact the American Mayor's role in the system may easily be exaggerated. In cities with the Commission or Manager form of government (just under half of cities over 5,000) the office of Mayor is of little account. In about a third of Mayor-Council cities, the Mayor is quite weak, judged by the absence of a veto power (Municipal Year Book, 1968, 105). This leaves about a third of all cities over 5,000 with a strong Mayor-Council form; and among larger cities (over 500,000) about three-quarters have 'strong' Mayors. Thus, viewed comparatively, the political significance of the Mayor is a characteristic feature of American city politics.

Formal authority of the Mayor

Formally, the role of the Mayor depends on his power to veto legislation, and to determine budgetary provision, and his powers of appointment and removal. In the bigger cities his formal powers are modified by his personal skills and the political scope available to him for enlarging or minimizing the opportunities of direct election and a four year term.

The veto power, provided in two thirds of Mayor-Council cities, is significant as a lever in the Mayor's relations with his council, giving him an equal power to deny legislation to them as they to him. Formal power to propose legislation is of less importance, since it cannot in itself amount to a right to make laws. The right to present financial legislation (the budget) is however important in mayoral power, for it is often difficult for the council to challenge a complicated budget once it is prepared. In some cases the council can only decrease or strike out items in the budget, and its overall shape if not its precise size is determined by the Mayor and his financial controller. If the latter is a running mate or an appointee of the Mayor, the Mayor's financial authority is substantial. In matters of appointment and removal the strongest Mayors (notably in New York and Detroit) have a wide discretion, but a constitutional provision for the approval of appointments by the council is common.

Thus city charters follow the political pattern of the state and federal government in providing for checks on the power of the chief executive. Generally the Mayoralty as an institution is distinct from the council (although the Mayor may be a member of the council) and he must develop his leadership role in co-operation or in competition with it.

The assembling of power

There are five areas in which the Mayor can assemble the power to govern. First, he has to enlist the support of the bureaucracies. Some have a measure of independence arising from separate election or appointment. But even where the Mayor can hire and fire, the professional administrator or planner has his own resources of integrity and expertise, and a solid tradition of hostility to political interference in the domain of the expert.

Second, the Mayor may gather force from his party. This is possible when the Mayor has a party of which he is the agreed candidate, with a coherent majority in council, and with a prospect of delivering rewards (jobs or policies) to party officials and supporters. These conditions are not consistently met. The Mayor may be the nominee of a group or a faction within the party; without dependable support in the council; and at odds with the party leadership in County and State; with few jobs to distribute and little chance of producing policies on which to build fresh support. It does not follow, however, that a divided and inefficient party always weakens the Mayor. The Mayor may find in such a party power enough for his support needs, but insufficient to constrain him. Mayor Wagner of New York City seems to have been in this position for some part of his tenure of office.

Third, the Mayor may look to the civic groups for the basis of a coalition capable of delivering both influence in the community and votes. Few Mayors arrive in office without backing of this kind, at least sufficient to finance the electoral campaign. One Mayoral candidate in Boston in 1959 is reported to have spent $500,000 (Levin, 1965, 23). The difficulty here is to make such groups the source of strength and not of obligations.

Fourth, the Mayor must establish good working relations with state and federal governments; with the Governor, and with the City's delegation in the state legislature; with the City's representatives in Congress, and with the federal departments. Here are opportunities for big city Mayors. Thus Mayor La Guardia of New York, lacking a firm party base in the city, secured the favour of both Governor and President; Mayor Daley of Chicago was able to press his views on the Office of Economic Opportunity (he was hostile to that Office's enthusiasm for the representation of the poor) through the Illinois Democratic delegation in Congress.

Finally, the Mayor can appeal to his personal constituency, the public. Success for Mayors, in staying in office and solving problems, seems related to skill in maintaining a rapport with the public. Popular support appears to be a political resource at all times, and one which is available more easily to the Mayor than to any other element in city politics. A skilful Mayor like Richard Lee of New Haven made urban redevelopment popular; Mayor Tucker of St Louis made good government palatable, even exciting, by his use of television (Greer, 1963, 194). But the public is naturally passive—it is unlikely to be deeply and persistently moved except by vast corruption or simple inefficiency, like a failure to collect the garbage.

Even a strong Mayor has to work for power; and every area in which he may develop power is also an area where he may lose it. The Mayor may choose a negative role, conceding to the most consistent pressures, a chairman and mediator at best; or he may adopt a conservative role, maintaining the community much as it is, arbitrating or moderating the pressures of special groups, and keeping himself and his associates in office; alternatively he may choose a leadership role, aiming in some areas at least to initiate change, development and reform. For those who have chosen the positive role there are different styles of leadership.

Styles of Mayoral leadership

Some small-town Mayors are part-time officials, combining

local politics with business or the practice of the law. Earlier even mayors of big cities sought other responsibilities—some were US Congressmen. Now most mayors are full-time politicians, and some, especially in the bigger cities, know no other profession, and expect, even after electoral defeat, to remain in politics. Many contemporary mayors writes C. R. Adrian (Adrian, 1961, 23) 'are amiable mediocrities, lacking in ability and imagination'. But the pressures of big city politics produce stronger candidates, who, while not representative of the office of Mayor, illustrate its potential and its limits.

(i) La Guardia of New York City

Fiorello La Guardia, Mayor of New York 1933-45, was a popular reformer, or radical demagogue. He was the first Mayor of New York drawn from the ranks of recent immigrants, and personally represented Italian, Jewish and Catholic elements. He was elected by a small plurality in a three-cornered fight, as representative of a Fusion party. His programme, apart from opposition to the Democratic machine, included non-partisan administration, merit appointments, Charter reform and major improvements in municipal services. Thus, it combined the traditional reform objectives with proposals adapted to the needs of the city in the depression.

A rule of American city politics is that Reformers are rarely given a second term. La Guardia's great achievement was to break the rule and stay in office for twelve years, being twice re-elected, and to carry through at least a part of the changes he talked of. Much of this programme was attempted and some achieved. The merit system was extended; several City Departments were reformed; relief and public works were provided, mainly with federal funds; the rapid transit system was improved and major air transport facilities (including La Guardia airfield) established; a new City charter was drawn up and operated from the beginning of 1938. Of course the City's major problems remained. Transport facilities were still inadequate, education suffered from some heavy-handed intervention, bankruptcy was staved off only by federal aid.

La Guardia's political position like that of all Mayors of New York was highly vulnerable. He had little influence in the Board of Aldermen (the City Council) and chose to ignore it. Tammany (i.e. the New York County or Manhattan Democratic organization) at first retained the Presidency of the Borough of Manhattan, but La Guardia contrived to manage the Board of Estimate and captured it in the elections of 1937. This body, which is much more important than the City Council, includes the Presidents of the five boroughs of New York and has considerable legislative and budgetary power. The Mayor needs the support of his Controller and President of the Council to outvote the Borough Presidents. (It is the Board of Estimate and the separatist tendencies of the Boroughs as well as sheer size which differentiate New York politically from other big cities.)

La Guardia had the advantage of support from outside the city. His relations with Governor Lehmann were good, and President Roosevelt, no friend to Tammany, supported him as a New Deal Mayor. La Guardia secured a generous flow of federal aid, under the Public Works Administration and other programmes, to build hospitals, a new subway and tunnel. But he failed to create an enduring party organization. The City Fusion Party of 1932 gradually waned. In 1937 the newly formed American Labor Party (an alliance of trade unions) supported La Guardia; but in 1943 the ALP coalesced with the Democrats, leaving the Fusion movement to fall apart, and the

Mayor with no visible means of support. La Guardia was too wayward and individual in style to be a party man.

Nevertheless his style was effective enough to sustain him through twelve years in office. He was not a demagogue, but he understood the politics of the streets. He would speak, or shout rather, in Spanish or Italian, 'vehement and declamatory, the picture of a militant crusader', pounding the rostrum, his voice breaking (Garrett, 1961, 110). He hated 'them', capitalists and industrialists mainly, but German Nazis too during the war, when his oratory was even less restrained. As an administrator he was quick and energetic but unwilling to explore facts or read papers; he was also difficult to work with, quick tempered and sometimes arbitrary.

His great talent was his energy, his readiness for a fight, ('you can't be a good fellow and a good Mayor'), his insistence on taking responsibility and on governing in the public eye. Like that other immigrant Lloyd George, whom he resembled in several particulars, he was a master of the art of getting things done. Looking at the organist in the spotlight in Radio City Music Hall, La Guardia said, 'You must keep both hands on the keyboard and both feet on the pedals and never let go!' (Garrett, 1961, 124).

This remarkable man was of course quite untypical of the general run of Big City Mayors. No other Mayor has read comics to the children over the radio (though Mayor Curley of Boston read poetry) or jumped up and down on a reporter's notebook, or vetoed the appointment of Bertrand Russell as Professor of Philosophy. Few personally attend fires, or lead raids on brothels, or consider themselves for the Presidency. Yet leaving aside his personal idiosyncrasies, La Guardia's style and career illustrate the coming together of two major strands in the tradition of the American City Mayor. One is the reformist. The other is more difficult to name; it is not exactly populist (which betokens rural radicalism) nor demagogue (because most of the time La Guardia's emotion and sensation were modified by reason and responsibility) but it included the visibility and communicativeness of the politics of the streets. La Guardia was a popular politician in a special sense, and a popular reformer. He succeeded in governing and even reforming an ungovernable city.

(ii) Mitchel and Lindsay

However, it has been possible to govern New York City without employing La Guardia's popular style. John P Mitchel (Mayor, 1913-17) was a manager, rather than a politician: 'the electorate admired but did not embrace him as one of their own' (Sayre and Kaufman, 1965, 692-3). He concerned himself mainly with administration, not legislation, and was unskilled in party relationships. Sayre and Kaufman conclude that 'under Mitchel the office of Mayor was a source of initiative, a centre of energy, a symbol of dignity and integrity' (694). They place him second only to La Guardia among the eleven Mayors from 1897 to 1953.

But Mitchel was not re-elected and John Lindsay, Mayor since 1965, had some difficulty in securing a second term. Lindsay has tried to be a WASP La Guardia, a liberal reformer, visible on the streets, but earnest, dignified, reasonable. Lindsay, like La Guardia, was elected as a Fusion candidate, but carried no-one with him into office. In a minority in the Council and the Board of Estimate, without a firm party base in the city, and faced with problems which are insoluble in the short term, Lindsay has been frustrated at every turn. He has still managed to develop the City's commitment to the poor, helped make the federal government aware of urban problems, and staved off bankruptcy. The city has avoided large-scale riots, but has suffered racial tensions and major strikes. In face of all this, Lindsay has proffered reasonableness, concern, trust, an urban Kennedy, perhaps too urbane for his tasks. It is evident after four years that in the big cities this version of the Kennedy style is simply not enough.

(iii) Mayor Daley of Chicago

In the beginning the style of American city politics is a front-porch style: '. . . one went over and sat on the committeeman's front porch and explained to him the need for a traffic signal or a sewer extension, and a few days later the facility was installed' (Reichley, 1959, 17). But these times have gone. The problems are more urgent and mostly insoluble; political communication is difficult. Mitchel's response was administrative; La Guardia's histrionic; Lindsay's ceremonial (shirt sleeves in the streets). Richard Daley, Mayor of Chicago since 1953, belongs to another tradition, the machine-based party. He is himself a party boss, but none the less a Mayor with good government to his credit.

'The Chicago area from a *purely formal standpoint,* can hardly be said to have a government at all' (Banfield, 1961, 235). There is separation of powers. The executive itself it divided, with directly elected Mayor, Treasurer and Clerk, and major policy areas in the hands of semi-autonomous boards. The Council is large and ward-based, so that every Councillor has his own source of power. Cook County, of which the city forms the major part, is a collection, not an organization. 'Every outcome must therefore be an elaborate compromise if not a stalemate' (p. 235). A single actor like the Mayor of Chicago 'can pursue a course of action only insofar as the formal decentralization is somehow overcome by informal centralization' (p. 237).

In this fragmented situation, Mayor Daley has built the power to govern through party organization. He has been Chairman of the Cook County Democratic Committee, controlling nominations and dispensing patronage which is still reckoned to include 6-10,000 jobs, as well as contracts. He has virtually controlled the city's delegation to the Illinois state legislature, about a third of which represents Chicago. There are of course limits to the Mayor's power. He needs some kind of co-operation from the Governor, who may be a Republican (Daley at one time had an agreement on spheres of influence). He must avoid outraging the public, his electors, though between elections the electors need the services of the Mayor, while he does not need their votes. Daley's political position has never approached that of an autocrat or a despot. On important issues he has had to conciliate interests, work along with professionals, accept overruling by the federal government (e.g. on the Fort Dearborn project examined in Banfield, 1961, 152-8). In other matters he has sometimes chosen to avoid personal commitment, e.g. in housing (see Rossi and Dentler, 1961, 249-51).

Daley is a boss but not a dictator. The power he acquired by tight party organization based on rewards has been used for 'good government'. In the context of American city politics this implies the reform (diminution) of corruption, efficient public administration, 'progressive' housing and welfare policies, and the institution of major projects to boost the city and its Mayor. Within this tradition, Daley has to his credit: a tight control of corruption and vice; a more efficient budgetary system controlled by the executive; an extended merit system; improved planning and zoning procedures; some urban renewal projects and an enlargement of transport facilities

(Banfield, 1961, 246-7 and Wilson, 1962, 72-3). Daley's notion of good government falls short of liberalizing the party or radically reforming the police. Thus his least attractive characteristics were displayed at the Democratic Convention of August 1968.

There is a good case for judging that Chicago has been better governed under Boss Daley than under his predecessor, Kennelly, who was not a boss. Certainly Daley has had sufficient power to govern and the costs of creating that power have not been intolerable. First there has obtained a politics of rewards, but far short of the undirected corruption which prevailed in the time of, say, James J. Walker, New York's Tammany-controlled Mayor of the 1920's. Second, there has been a diminution in the vigour and meaning of electoral politics; but the machine has dominated mainly in the primaries. Daley's capacity to deliver votes has been doubted with some justification by recent Presidential candidates.

(iv) Richard Lee of New Haven

Richard Lee, Mayor of New Haven since 1953, offers another type of Mayor, and one perhaps nearer to the liberal ideal—a positive politician-administrator. He was positive in the sense of coming to office with a programme which he spent most of his efforts to fulfil; in Duane Lockard's typology, Lee was a programme-politician (Lockard, 1961, 41). Lee was 'the first Mayor in the country to have made urban renewal the cornerstone of his city's administration as well as of his own political career' (Lowe, 1967, 405). This did not look like a workable strategy, even in a city of 150,000 people. Urban renewal projects notoriously antagonized many people, dispossessed tenants and house-owners, disadvantaged businessmen, disappointed contractors, envious politicians. Nor does the New Haven Charter give the Mayor invulnerability or power to execute a project without support. Executive functions are shared with semi-autonomous boards and with directly elected officials, while legislative and budgetary functions lie with the ward-based Board of Aldermen.

Despite the obstacles, Lee's Mayoralty has succeeded in that he maintained himself in office, and accomplished a substantial part of his programme. The serious riots in New Haven in 1967 indicate the limits of his achievement, but do not indicate overall failure. Lee's success lay in his capacity to exploit the potential of his office, to develop and apply power in a selected arena.

He was first a party leader as well as a Mayor, one of a triumvirate with the Democratic Boss (John Golden) and the Director of Public Works. He used his ample resources of appointments, contracts, honours, favours, and he wooed his Aldermen, for example, providing their wards with highly visible improvements like playgrounds. As party leader, Lee was both boss and prisoner; his party supported him, but not without the exertion of leadership on his part and occasionally a veto on their's (as for example when he attempted governmental reforms which would have enhanced his own independence). Mayor Lee also saw the value of experts and planners, and learned how to use their talents without losing his own power and popularity. His Development Administrator, Edward Logue, was reputedly one of the best in the country. He enlisted the sympathy and co-operation of private developers; and had good fortune in finding one at least who followed civic as well as business values. He had good fortune too in securing massive federal aid—more per capita than any other city.

Above all Richard Lee worked unceasingly to generate popular support. A former public relations officer (for

Yale) himself, he used the press with high skill. Those playgrounds always had a grand civic opening with a large sign reading 'An Awakened New Haven Builds for the Future'; Lee could be seen in the cab of the wrecker's truck, ready to knock down slums and 'even the rats, symbols of slum evil, had a press conference' (Lowe, 1967, 424-6). However, Lee's concern for popular approval went beyond such attention to the media. He promoted a Citizens Action Committee which involved several hundred people in the promotion of urban renewal projects. Richard Lee was a tough and powerful Mayor, driving himself and his city very hard. But his over-riding strategy was to rely on persuasion rather than threats (Dahl, 1961, 119). In consequence he survived and prospered, one of the best modern examples of the programme-politician-administrator.

The Mayor and political leadership

These examples—La Guardia the popular reformer, Daley the boss, Lee the programme politician—illustrate the potential of the office of Mayor in some big cities. There are others of the same type; but there are also many who fill two of Lockard's less admirable categories, Evader and Stooge (Lockard, 1963, 41) and a long tail of 'amiable mediocrities'. Even for strong and positive Mayors in big cities, power is discontinuous, varying with the issues and the ethos of the times. Mayor Walker was a product of the 20's as La Guardia of the 30's and the New Deal; Daley's party system is affected by a change in the Presidency.

Nevertheless, most Mayors starting with strong Mayor Charters have some choice in what they make of their office. Their choice depends both on their predilections and their capacities. They may choose to be Clerks or Leaders. Power is available, but is rarely untrammelled.

Leadership requires the building of support from party, or public or a coalition of groups; but support carries with it commitments and obligations. Similarly, the Mayor becomes naturally the personalization of city government. 'The Mayor (of New York) and his office are the visible expression of the city, its personification as an organized community . . . spokesman for the city's largest constituencies' (Sayre and Kaufman, 1965, 657, 698). But personification brings personal vulnerability: every time the passenger pays 25 cents for his fare he thinks of the Mayor. Mayor Lindsay has been particularly unlucky in having to accept responsibility for numerous strikes of major utilities, two power failures, and an unusually severe blizzard. The danger is that the Mayor inherits responsibilities unmatched by his powers. In this sense the Mayor may reign but not rule.

The Mayoralty, like the Presidency, is crucial for leadership. This was true in New Haven: only the two Mayors (Celentano and Lee) 'exerted direct influence in all three issue-areas studied' (Dahl, 1961, 181). It was true also in Philadelphia, according to groups like the City Planning Commission and the Delaware River Authority (Reichley, 1959, 119). Other studies show that Mayoral leadership is more likely to secure federal aid, and more likely to secure fluoridation of water supplies (the latter a litmus test of reform against obscurantism).

The choice for mayors between clerkship and leadership is not of course as simple as this analysis may imply. The circumstances and traditions of local politics sometimes prohibit positive leadership. Personal style and capacity may not suit the political opportunities. Finally, a Mayor may prefer the negative and mediating role. In that choice he would be holding to one of the enduring values of American political tradition.

(ii) The Boss

Origins of machine politics—the old style boss

A boss is a professional party-organizer, occupying a base in city government, and working by means of rewards and services for voters on the one hand and on the other for a clientele of businessmen, community interests and the community at large. The classic period of boss and party machine was in the last decades of the nineteenth century. American cities were then expanding very rapidly under the impact of industrialization and of massive immigration. Such conditions produced urgent needs: the needs of the immigrant poor for jobs, food, welfare; of businessmen for unrestricted opportunities; of utility operators for franchises, of saloon-keepers for tolerant regulation. These needs could have been neglected, or met in other ways. The party machine was a characteristically American response, arising from the wide distribution of the vote and a free-booting, entrepreneurial tradition of politics. 'What are we here for except the offices', said a Convention delegate of 1880 quoted by Bryce. Bosses, Bryce commented, were 'the offspring of a system. Their morality is that of their surroundings. They see a door open to wealth and power, and they walk in' (Bryce, 1888, 201-2).

In this period, machines dominated politics in many big cities, and achieved unprecedented standards of corruption. In New York City, Boss Tweed and his friends raised the City's bonded debt by about 40 million dollars in the two and a half years ending with Tweed's fall in September 1871. Corruption on this colossal scale aroused a Reform movement signalled by Lincoln Steffens' famous exposure of 'The Shame of the Cities' (published in 1904). Thus, fortuitously city politics acquired a tradition which at least permitted a bastard form of two-party dialogue and alternation. But the boss and his machine are more significant and more complex political phenomena than the great nineteenth century exemplars can demonstrate; in particular, a boss is not necessarily either corrupt, tyrannical, unjust or undemocratic.

The machine was based on small electoral units (wards, precints, districts), organized by personal contact for an electoral system which included frequent elections and long ballots. The machine controlled the city administration and the courts, and financed itself from the sale of nominations and office, a levy on office holders, contributions from interested businessmen and direct subsidies from the City Treasury. Jobs were its chief stock-in-trade. Boss Kelly of Chicago, it was said, 'walked around with 9,000 jobs stuck in his back pocket' (Meyerson and Banfield, 1964, 79). James Curley of Boston boasted that he disposed of 60 jobs in his first two hours back in his office after a period in gaol (Curley, 1957, 335). In addition to jobs, the Boss had in his gift contracts, favours, privileges and protection. Much of this was corrupt in that the Boss was concerned not with efficient employees, economical contracts, a just police force, but simply with gratitude and support. These were the objectives also of his services to the poor: but at least the services were genuine enough. The machine professionals' 'task of bridging the gulf between the individual and his government, or at least of seeming to do so, forces on them endless small duties which take them to all corners of the city, at all hours of the day and night' (Reichley, 1959, 103). The duties included food and clothing, coal, Christmas and Passover baskets, picnics; family advice and attendance at funerals—'Help, you understand, none of your law and justice, but help' (Martin Lomasney quoted by Steffens). 'Bread cast on the waters returns in the shape of votes'. 'The poor,' said G. W. Plunkitt of Tammany Hall, 'are the most grateful people in the world'.

The Boss was, and is, a professional, engaged in politics to make a living. As Bryce noted, his interest in political questions may be quite secondary. Formerly politics provided 'an escalator for self-advancement' (Garrett, 1961, 4). Professional advance was not limited to politics. Businessmen and lawyers in particular prospered in their own trades by attention to politics, and politicians acquired successful careers on the side, for example in insurance and contracting (Gosnell, 1968, 49 and 67). Curley had some major business successes. But bosses, like true professionals, seem to have liked their work, savouring the power and status it gave them, and enjoying the pursuit of the great game of politics.

The famous bosses of American urban history form a curious and striking gallery of portraits. James Curley of Boston, four times Mayor of Boston, once Governor of Massachusetts, and twice in gaol, is the best example of the player of the great game. His book (Curley, 1957) is perhaps too self-consciously part of the mythology which has made Bosses into Robin Hoods, Czars and Father Christmas. His high skill as a political tactician fell often into deviousness and sometimes into self-delusion. But his sentimental charitableness does not hide the toughness, the contempt for reform (the 'Goo-Goo's'), the assumption that support is open to purchase, and the ruthlessness ('Do others or they will do you'). Richard Croker of Tammany was asked during the Mazet hearings of 1900, 'Then you are working for your own pocket, are you not?' He replied: 'All the time, the same as you' (quoted in Lowi, 1964, 84).

A boss was often an immigrant himself, drawing on ethnic groups for support; making no nice bourgeois distinction between gambling or prostitution and honest trade; and, like Curley, and Hughey McLaughlin of the Brooklyn

Navy Yard, 'Big and strong, handy with his fists in a fight' (Lowi, 1964, 35). Irishmen were prominent among the earlier bosses, not because of Irishness, but because they were an immigrant group sufficiently large and cohesive to seek to run their own communities. Boss Cermak of Chicago was a Czech, autocratic in his control of patronage, unduly close in his connections with gambling interests, and mismanaging his relations with the President and national party. Ironically in 1933 while trying to mend his relations with Roosevelt and Farley, he was killed by an assassin's bullet intended for the President. David Ryan, a later Chicago boss used patronage and little else. Where the modern public administrator spends much of his time reading and writing, Ryan only talked, and that not much. 'He did not even own a brief case.' (Banfield, 1961, 24.)

The boss as party manager

The modern boss might better be called a party manager to distinguish him from his illustrious but corrupt predecessors. The transition can be seen in the new image developed by Carmine de Sapio as Tammany leader in the 1950's. The Italians had arrived too late to enjoy the triumphs of the Irish. In de Sapio's time half of the Democratic leaders in New York City were college educated, and de Sapio himself studied English, wore sober suits, published an article in a scholarly journal and evinced liberal sentiments—though some of his supporters showed a traditional addiction to honest graft and gerrymandering (Garrett, 1961, 327). But 'bosses' like John Golden of New Haven (Lee's collaborator) and Congressman Green of Philadelphia are simply party managers, who, patronage apart, would be recognizable in some party-controlled English Boroughs.

Few bosses, now or earlier, have been absolute autocrats in the party, still less in the city. Some have had to share power with ward-leaders, and the Boss has been, so to speak, put into commission as a collection of Chieftains. This was the case in Chicago before Daley. Sometimes rival machines have operated within the party. Where the boss has chosen to occupy a party not a city office (a common arrangement), municipal authority has checked party influence; but of course, where the two have combined, as with Daley, the boss/mayor has been powerful indeed. The machine's control of the vote has been tight in some wards, amounting to a capacity to deliver a vote, or even to switch it from one party to another. This has operated most consistently in primaries, which are important enough, being the only significant election in areas dominated by one party. But lapses occur, and the machine itself may generate opposition. A better educated electorate is normally more independent in its voting behaviour. Moreover, the scope for electoral malpractice has diminished, through publicity and a changing climate of opinion.

The limits of the power of the Boss show in the fluctuating influence of the machines in the cities. Boss Croker, it is said, once governed New York City from his estates in England (Lowi, 1964, 175). But since the new Charter of 1897 Tammany has had only fitful control of the Mayoralty: there is no question of a monopoly. Boss Curley experienced more defeats than victories, and Daley has occasionally lost local elections. The machine must compete in politics and, like Green in Philadelphia, may have to bargain with a successful reform movement (Banfield, 1966, 119-20). Gosnell wrote in the 1930's; 'Party bosses in Chicago have been unscrupulous in the use of public office to build up their political organizations; but from time to time they have been compelled to recognize the larger community interests' (Gosnell, 1968, 76). Republican machines have operated in some cities, Cleveland for example; and in others, like Minneapolis, party politics has been loosely organized and a ward-based, patronage-dispensing machine has not developed (Morland, 1960, 283-6).

The decline of the boss?

The decline of the Boss is attributed to the ending of immigration (the flood was over by 1920) and the consequent drying up of the machine's best source of customers. Second, in the New Deal period the federal government moved into the business of social welfare and Boss/President Franklin D. Roosevelt served the underdog and captured his vote. Both these explanations are valid though subject to qualification. Immigrant (new and unassimilated) families remained and were still open to political organization : the only new factor was the absence of the immigrant just off the boat, always a customer ripe for Tammany's attention in the port of New York. The inauguration of federal welfare was a more serious blow to the machine. True, New Deal schemes like the Works Progress Administration provided a fresh source of patronage; but this was exercised from Washington, with Senators and Congressmen as agents, and subject to Presidential favour. (The New York Democratic organization has suffered from Presidential disfavour under both Roosevelt and J. F. Kennedy.) Moreover, in the end, federal welfare and a developing economy have raised the poor beyond the services of the machine. Worse, the newly prosperous workers have started a new immigration to the suburbs, thus splitting and scattering the classic constituency of the Boss. A suburban middle class boss is not a boss at all—but a party manager. 'What beat (de Sapio) in the end was the middle class . . .' (Murray Kempton in the *New York Post* of June 9th, 1960, quoted in Wilson, 1962, 642; see also Wood, 1958, 169-70).

There are other explanations for the weakening of the machines. The Reform tradition persists and its periodic upsurge of anti-party legislation and anti-machine, 'throw the rascals out' sentiment, is strengthened by the teachings of modern public administration. Appointments should be not only for merit but for job-oriented training; financial management is a matter not simply of avoiding open corruption and waste but of positively seeking regulated and audited efficiency; community welfare is a task for professional case-workers; communication, a job for counsellors in neighbourhood city halls. Moreover the machine worked on a continuous belt. Power produced votes and rewards, which produced power; once the belt broke, the machine was difficult to restart.

However when all the explanations for the decline of the machine have been reviewed, the fact remains that machines still operate. American society is no longer so responsive to the rewards and controls of the machine boss. But the politics of jobs and rewards, advice and welfare still obtain in some cities especially where there is a large low-income population (see Frost, 1961 and Watson, 1963, 22) City machines reappear after years of quiescence. Even Tammany Hall, no longer an actual building, continues to operate, though as a neighbourhood rather than a city-wide force. Party organizations still function, and the city machine is part of a state and national party organization. Parties still need money and support; and its clients still want rewards and services. The style of politics persists, though the organization has been changed by the great secular movements of society and politics—the waning of immigration and the coming and passing of the New Deal.

Machine politics appraised

Boss rule plainly had major defects. Its basic method led to corruption and inefficiency. It was concerned only with maintaining itself, and was indifferent to politics in the sense of policies for problem-solving and community development and the public agitation of issues. Spontaneous political organization was discouraged (see Gosnell, 1968, 83-4) and reformist tendencies were heresy. A New York reform leader once naïvely asked for posters for the Stevenson campaign (after all this was the Democratic Party?); 'those political hacks . . . took me aside and told me that our man was Dan Fink, who was running for judge of the municipal court . . .' (quoted in Wilson, 1962, 61). Boss rule excluded the possibility of dynamic politics and an extended scope for city government.

Yet, there is a good deal on the credit side. 'From the standpoint of politics the great mass of the people are interested in only three things: food, clothing and shelter' (Martin Lomasney quoted in Curley, 1957, 157). These the machine endeavoured to provide, deriving some satisfaction from Steffens' description of boss-ridden Philadelphia as corrupt but contented. The gains were not limited to the material. The machine was a medium of communication well suited to the needs of the immigrants and the poor: 'the only true means that we have yet developed to communicate to the seats of the mighty in government the sentiments of the great mass of citizens and individuals . . . and to give these sentiments a genuine role in the formulation of policy' (Reichley, 1959, 107).

Thus the machine related power to the people, it helped to incorporate new groups into the polity. It took society as it was, provided for its needs, harmonized its pressures and created a community. Politically the machine brought together the power by which the conflicts of a decentralized system could be managed, and a socially explosive situation defused. In Chicago, for example, 'when the Mayor ceases to be a boss, he will not have the power to run the city as it should be run' (Banfield, 1961, 258). Further, and through no deliberate effort on its own part, the machine produced a counter-movement, a reformist opposition, and thus brought about, sporadically it is true, a competitive form of urban politics.

All this, it must be admitted, is to emphasize the brighter possibilities of the machine: to be fair, it has often been judged by its worst features. Moreover the alternative to boss-rule was often not dynamic reformist politics and good government. The alternative was more likely to be rule by demagogue, or by interest groups, or by the middle classes. These are still possibilities in the contemporary American city, though in historical circumstances radically different from those which first gave rise to the machine.

(iii) The Manager

The Manager plan

The city manager is an American political invention which has had a remarkable impact. It has been adopted in just over half of cities in the medium range of 25-250,000 population (Municipal Yearbook, 1968, 53). However, only a few big cities have adopted the plan (among them, Cincinnati, Dallas, Kansas City, San Diego) and some like New York and Philadelphia have deliberately rejected it.

The city manager plan, as developed by R. S. Childs, was based on the unification of power in a small, elected and politically responsible Council, which delegated administrative authority to an appointed manager. An essential element in the early design of the system was the short and non-partisan ballot, with no separately elected Mayor (though a strong Mayor was not ruled out). A Council small enough to be a Commission was favoured; so too was proportional representation. The plan was clearly intended to banish the demon of party from City politics, replacing it with purity, efficiency and democracy. Some of its supporters made sweeping and naïve claims: the plan was 'the most perfect expression which the American people have yet evolved of the need for combining efficient administration with adequate popular control'. 'What the whole world is witnessing is the emergence of government by experts' (L. D. White quoted in Banfield Reader, 1961, 251-2). Yet, for all their naïve optimism and faith in mechanistic adjustment, the reformers' combination of old constitutionalism and new reformism had created a fundamentally revolutionary scheme which abandoned the traditional constitutional device of separation of legislative and executive powers.

In practice, the manager plan worked well only when manager and council were in agreement. Otherwise the unified powers fell apart and defied either reunification or coherent redistribution. But these problems of power were not at first perceived and the plan was promoted with a characteristically American enthusiasm. A National Conference on Good Government of 1894 led to the establishment next year of the National Municipal League. Supporters of what must properly be described as a reform movement included Charles Beard and Woodrow Wilson, both distinguished academics. The International City Managers Association was founded in 1914 and began the vigorous propagation of the plan which sustained its members. A good deal of the literature on the city manager is recommendatory rather than analytic. The plan spread rapidly from Staunton (Virginia) in 1908, to 450 other cities by 1938. Generally it was adopted in cities influenced be reform movements (like Cincinnati) and in many small cities with a low intensity of politics. But of course it was promoted and adopted in some cases for reasons unconnected with reform. In San Diego its establishment was caught up in a municipal dispute with the fireman's union; in Dallas the plan rose with the support of the Dallas *News*, and almost fell when the *News* came to disapprove of the first City Manager. By 1968 the manager plan had spread to over 1,200 cities over 5,000 population. It has proved particularly popular in California.

The average City Manager works in a small town: over half in towns under 25,000. Even so he is likely to be college-educated and a newcomer to the town; and he will move on within a few years. In the small town his main problems lie not in power relationships, but in multiplicity of functions. His attention to finance or planning is more likely to be disturbed by a breakdown in the waterworks or the electrical plant than by a breakdown in political support. In the bigger cities however, the city manager is involved in political relationships similar to those of the Mayors, Bosses, and party and group leaders. When the manager tries to manage in a situation of tension and disagreement, the political flaw in the manager plan is revealed.

The Manager and policy-making

The original plan was based on what were thought to be fundamental elements in the British parliamentary system: the distinction between policy-making and administration, and the unification of both, with responsibility to the legislature i.e. the Council. Strictly, the one precludes the other, but the confusion is British too. Thus the city manager was either an administrator but not a policy maker; or he was both at once, but in a manner which was private and inconspicuous, neither visible nor directly responsible to the public. However, neither of these patterns proved to be consistently practicable. At the levels of government in which the manager was concerned it is difficult to distinguish objectively between policy and administration. Neutrality in values is notoriously a value-laden position. The simplest test in practice was the reaction of the citizens: decisions to which they were indifferent were value-free, hence purely administrative, and appropriate for a city manager endeavouring to avoid the political. But in a situation of public agitation demanding positive action, this theory would make of the manager a passive clerk. Similarly the inconspicuous manager was likely to be impotent in the face of urgent problems, since he was prevented by his role from endeavouring to secure agreement and ease conflict over policy.

Perceptive supporters of the manager-plan have devoted much thought to these problems, and the ICMA's Code of Ethics for managers has been debated and amended with all the pious enthusiasm of the early church revising its creeds. The 1924 version of the Code stated bluntly: 'No City Manager Should Take an Active Part in Politics.' Later a distinction was attempted between community leadership and political leadership—the Manager was still (in the 1938 Code) 'in no sense a political leader'. But, according to the gloss of a leading authority on the plan, community leadership might include 'explaining the work of city government, and preparing new policies for it' (which might be by public speeches); 'negotiation with private citizens and community organizations' to secure support; 'speaking in a campaign for a referendum on a bond issue, if the question at stake is not identified with factional interests or a factional disagreement'. However, community leadership 'most certainly excludes an appeal by the manager to the voters over the heads of the councilmen'. (These quotations are from Stone, 1940, 243, which, with its companion research studies is sympathetic to the manager plan but scholarly and perceptive.)

This conception of community leadership concentrates on the open public activities of the manager. The question remains of his relation to the Council in the making of policy. Here the manager's role lies in the development of policy, providing facts and advice and making recommendations. This role is accepted by Stone and his colleagues as policy leadership: 'it is generally impossible for a city-manager to escape being a leader in matters of policy, for it is an essential part of his administrative job to make recommendations' (Stone, 1940, 243). The Revized Code of 1952 also recognized the function of policy advice, but endeavoured to set limits to the manager's part in communicating policy to the public. The manager should not defend policy in public until it is adopted by the Council; he should avoid 'public conflict with the Council on controversial issues'; and must allow the Council credit for the establishment of policies (Articles 4 and 5 of Code of 1952).

These carefully devised guide-lines have sometimes proved unhelpful. The city-manager's profession has developed a positive and innovating style and problem-solving goals. In one study (Ridley, 1958) all 88 managers agreed that they had a 'definite responsibility to participate in policy' and that they were expected to propose ideas on policy to the Council. Most of them (77 of 88) stated that they 'as a matter of course initiated policy'. In a study of the small town of Oberlin (8,000) the City Manager was rated by the author high in influence in three of the six conflicts reviewed (Wildavsky, 1964, chapter 18).

Moreover, if city managers are influential in policy-making, their influence with the Council is unlikely to remain a purely private relationship. Stone admits: 'The city manager's recommendation on an important policy, even if he makes it in an executive session of the Council, is usually a matter of common knowledge' (Stone, 1940, 243). Thus tension and conflict may arise from the relationships involved in the manager plan. Then differences in policy and political attitudes are intensified and embittered by distrust of the system and of its innocent protagonists (and victims). Councilmen readily see their managers as 'Blunt, bull-headed, inflexible, cold and unfeeling, proud and self-satisfied, stubborn, red-tape happy, . . . arrogant, conceited'. Meanwhile the manager is inclined to criticize Councilmen as 'easily influenced, too lenient, too friendly, touchy, and sometimes sceptical and hard to impress' (Carrell, 1962, 204-5). Thus the city manager, like the Mayor, sometimes got the worst of both worlds, with high vulnerability and inadequate power.

Assembling power

Like the Mayor, the Manager must often look to politics, not to Charters and Codes, to solve the problem of assembling adequate power to manage his city. He is not without political resources to assist him in securing and maintaining the support of council and citizens. He has, or should have, managerial skills in complex matters like finance, planning and intergovernmental relations; he can make his office indispensable for fact-finding, communication and conciliation. The municipal bureaucracy itself provides the manager with a source of support, both within City Hall and as a voting group in the city; though there are cases of departments, notably police departments, conducting themselves as independent and even rival empires. The manager may also derive some respect and influence from a professional tradition and his own professional standing. His tenure may be more soundly based than elective offices. However, there is fairly rapid turnover of managers: in 1966 about 10% moved other than by retirement or death. Probably such mobility reflects American attitudes to the profession of government rather than the flight of managers defeated or defiant in political conflict.

In promoting his policies with a sceptical or indifferent Council, the manager may employ several strategies. The most natural, and least costly, is depoliticization and the professionalization of issues. The epic seven-year-battle between the City Manager of Cambridge, Massachusetts and his Council (Stein, 1952, 573-620) was fought by the Manager largely on the grounds that personnel matters were an area for his professional prerogative. Certainly there is now a substantial body of municipal expertise in financial, planning and personnel fields as well as in the traditionally technical affairs of water engineering and the like. Two managers were quoted in 1959 as saying 'What were once flaming municipal issues. . . . have in many instances become routine elements of city administration' (Harrell and Weidford, 1959).

There are greater risks for the manager in seeking to organize public support. Appeals to the public directly or through the press are open to challenge by councilmen, and are by no means assured of success. The encouragement or even covert promotion of citizens' groups is safer. In Cambridge, the Cambridge Civic Association provided effective support for the beleaguered Manager; but the CCA was not manager-promoted and proved to be an independent force in city politics. Some citizen groups of this kind are in fact fronts for city interests. The Manager who exclaimed: 'God bless all civic associations, they are the city manager's ward machines' (Wood, 1958, 184) perhaps overrated their utility.

In the end the manager has to win the support of the council. Here a favoured strategy is for the manager to develop and propose a range of options from which the council may choose. Mayors are on record as positively approving this arrangement which leaves initiation to the manager and gives the council the opportunity to 'knock down ideas' (McCorkle, 1952). This pattern is neatly illustrated in Adrian's tabulation of leadership in municipal issues in three Michigan cities. The administrators were pre-eminently proposers, the Councilmen operated a veto, though outside groups were also important in both categories (Adrian, 1958). This pattern is, of course, one version of the British system by which civil servants advise on policy; its effect on the distribution of influence varies with the participants but certainly allows managerial dominance.

Few Councils are likely to follow in detail Stone's classic advice: 'The function of the Council is not to be a check on the city manager. Its function is to enable the city manager, as head of the administration, to render the most effective service to the city—by legislation, political leadership and public relations work, by active study and advice, or sometimes by doing nothing at all' (Stone, 1940, 254). In consequence, the city manager must nurture carefully his relations with councilmen: he must be a skilled and sensitive politician. After four city-managers had gone through San Diego in the space of three years the fifth (with a reform movement to back him) said: 'To make a really permanent contribution (the Manager) has to move slowly, take account of political realities without being too pliable, and watch his personal relationships carefully. A great many managers have ruined themselves and their work by rushing in and grabbing a lot of red-hot pokers' (Stone, *Nine Cities*, 1940, 180).

Patterns of manager government

The range of manager-council relations runs from submission to dominance. The plan has tended to work in three main ways. First, the city-manager has been incorporated into a party-organized conciliar system as in Kansas City under the old Pendergast machine. The manager then prospers or falls with the council, and normally forfeits any claim to professional tenure. He is equivalent to the party-appointed high political executives who come and go in Washington with the President—an honourable and even admirable device, but not one envisaged by the first promoters of the Manager plan.

Second, the manager may become a clerk, without leadership aspirations. This is the case in many small towns, which are unlikely to be interested in innovation, and are not therefore seeking leadership. Again this might disappoint the founding fathers: but the stability of the city manager system in so many small towns suggests that the manager-as-clerk adequately meets the articulated needs of his community. Adrian in a study of three cities in Michigan (50-70,000 population) noted that only about two issues each a year could be regarded as important, judged by controversy, press-coverage and time required for solution (Adrian, 1958).

In the third pattern of manager government the manager fills the role intended by his progenitors, by the later ones at least: he strengthens elective political leadership with professional policy advice and efficient administration. If it is assumed that the larger cities have social and administrative problems requiring positive political leadership for their solution, then this third style of the manager plan is important for the good government of cities, and the most significant for the historian and political scientist. In what circumstances does this third pattern, 'true' city-management, arise? and what factors contribute to its persistence and effectiveness?

The answers would appear to lie much more in the political tradition or ethos of a community than in its socio-economic character. (But the matter is far from resolved.) It seems that city manager government tends to develop and succeed:

> where politics have not been dominated by a single party machine, in particular where non-partisan and at-large elections have obtained;
>
> where citizens and community leaders prefer positive, creative and centralized government to passive caretaker government;
>
> where council and city manager are agreed on these goals;
>
> and where the city-manager plan itself has been introduced without major controversy and is generally accepted.

In sum, the city-manager government has flourished in a reformist atmosphere, sympathetic to middle-class values. Without this background, even a highly skilled and experienced manager may fail.

Size of city does not seem to make much difference to the working of the plan, at least in the middle ranges of population. Williams and Adrian studied four cities of roughly equal size (50-70,000) and found striking differences in management. These they ascribed to 'the value structure of the community and institutionalization of political values' (Williams and Adrian, 1963, 308). Similarly, Stone's classic study concludes that the plan works best in 'community governed cities', i.e. those with strong ideals of community service (Stone, 1940, esp 9-11). However these conclusions do neglect one possibility—that plan and ethos developed in parallel. Certainly some city managers have had to build up a favourable ethos from unpromising beginnings, and plan and ethos have flourished or perished in a mutual relationship.

The Manager plan assessed

The city manager plan has not of course solved the problems, political or managerial, of American city government. These problems are more difficult than the early promoters of city-management envisaged. The analogy of city government and the business corporations was particularly misleading. As the San Diego Trades and Labour Council said: 'Corporations are for the making of dividends. A city government is charged with the welfare and happiness of human beings' (Stone, *Nine Cities*, 1940, 158). Inevitably urban social conflict is too profound and persistent for politics to be taken out of city government. The great political defects of the manager-plan are that it has discouraged popular participation and neglected the interests of the low-income groups. But this may now be changing, at least in comparison with machines, which have hitherto served the lower classes better (see Greenstone and Peterson, 1968, 267-292).

The Manager-plan has great achievements to its credit. It has moderated excessive partisanship, diminished corruption, concentrated authority and improved organization. It has raised the standards of professional city administration, introducing efficient budgetary procedures, professional planning and personnel techniques, and higher standards of public service. The present tendency is, nevertheless, for the reassertion of political leadership. Some supporters of the plan agreed in 1964 to permit directly elected Mayors, but without increased power. The election of other officers, Treasurers, Police Chiefs and the like has declined. By 1967 over half of manager cities had directly elected Mayors; and in those cities the Mayoral election tended to be front page news and the Mayor a significant public figure; while the managers tended to come and go more frequently than before. There is a tendency indeed for managers to become associated with directly elected Mayors and to rise and fall in effect with popular election.

In this way responsibility comes to be centralized in the Mayor and the Manager; and the manager-plan city converges in its politics with those strong Mayor-Council cities which have appointed Chief Administrative Officers. This signals a development rather than a defect of the city-manager scheme; a development which allows a running compromise not between administration and politics, but between professional efficiency and popular responsibility. This is precisely the aim of the earliest propagandists for the manager-plan.

(iv) The Administrator-Entrepreneur

The Administrator in city politics

City politics, like industry, has undergone a managerial revolution, and, with or without city-managers, administrators have acquired great influence. A boss might not even own a brief case: but the administrator owns and uses one, stuffing it with the reports he must read and the policy papers he has written. The functional differentiation of politician and administrator is not of course sharp or complete. The administrator's functions certainly run to advising on policy, and this may include the assessment of acceptability and action aimed to secure support.

There are many pressures on an administrator to overcome personal and systemic inertia—the demands of personal satisfactions and professional esteem, career expectations. Pressures arising within the office are modified by the persuasions of the administrator's clientele, and the policy goals set by his executive leader (the Mayor). This makes up a triangle of forces which could fix the administrator in immobility, or drive him strongly to one side. The administrator is thus an element in the community power structure, and certain forces and characteristics of American city politics tend to give weight to the administrative element.

First, the reform ethos, with its distrust of party politics, has given the administrator legitimacy and hence lowered his vulnerability. At the same time non-partisanship has inhibited the development of regular party organizations likely to strive for influence in some policy areas. Second, the power thus won from the politicians has often been deconcentrated in the cause of independence and specialization to semi-autonomous agencies and boards. Such agencies have often operated federally aided programmes, controlling by contacts and expertise independent channels of access to federal finance. These administrative units have sometimes been colonized and energized by groups with dynamic purposes. Finally, the professional, unlike his British counterpart, is not limited by the self-denying rectitude of a corporate profession; he is rather a freelance willing to sell his talents elsewhere. Altogether it has been possible for some top administrators to 'set the tone' of the Mayor's administration by 'increments of innovation at the centre' (Lowi, 1964, 5).

The normal pattern is, however, very far from bureaucratic tyranny. 'The modern city is now well run but ungoverned because it now comprises islands of functional power before which the modern mayor stands impoverished' (Lowi, in Foreword to Gosnell, 1968, X-*XI*). But the administrator is also impoverished by this lack of co-ordination: too many chieftains make for very petty chieftainries. Moreover co-ordination in a large city may prove difficult if not impossible to bring about, as is suggested by the recent experience of New York's Human Resources Administration. Again, the 'islands of functional power' have sometimes been subjugated rather than developed by invading groups. This is notably the case where organized municipal employees have acquired preponderance of power in particular agencies. Police and fire services provide obvious examples (e.g. see Sayre and Kaufman, 428-31).

It is significant that some administrators move into elective office, in search of power, often as running mates for Mayors they have served. But these are by no means representative examples. Career administrators persist and may, as petty chieftains, exercise a significant degree of influence. The largest class of influential administrators (apart from city managers), is the commissioner or department or agency head. There is also a small class of administrators occupying the recently developed office of General Manager or City Administrator. Apart from particular offices there is a style of administrator, characteristic of American cities, the administrator-entrepreneur.

Aides to the Mayor

In the last twenty years several large cities, including Boston, Newark, New Orleans, New York and Philadelphia, have established an office intended to provide high level administrative assistance and advice to the Mayor and known variously as city administrator, managing director or chief administrative officer. In Philadelphia the Mayor has four aides, directors of finance and personnel, a city representative (to 'cut the important ribbons') and, most powerful, a managing director (see Clark in Banfield Reader, 1961, 453-62).

Such officials are almost always appointed by the Mayor and act in theory as the Mayor's agent. The powers of the administrator vary. In Philadephia the managing director has powers of appointment and supervision of department heads. Both are of course subject to the Mayor's approval, but in the matter of departmental supervision Mayor Clark said, 'it's a rare day when I see any of those commissioners' (Banfield Reader, 1961, 453). There are, however separate directors for finance and personnel. In New York, the City Administrator survived an early contest with the Budget Director, but the office has not in fact developed as it was intended into a substantial support of the Mayor as chief executive officer. Instead his most impressive functions have been 'improving the organization and management practices of the different departments and agencies, advising on and recruiting candidates for appointment by the Mayor as Commissioners, resolving conflicts between and among departments and promoting better integration of interdepartmental activities affecting common problems' (Caraley, 1966, 16). In some cities the administrator is a member of a Management Cabinet. In American politics the term Cabinet normally implies at most an interdepartmental or co-ordinating committee and is not a body of elected and responsible party politicians as in Britain. Nevertheless, the administrative Cabinet is an instrument for the accumulation of administrators' influence.

The office of city administrator is plainly still in an ambivalent position open to further development. Mayors, like Presidents, need help, especially 'political help and advice, supplied by persons sensitive to and sensible about the Mayor's full dimensional role in the city's political contest' (Sayre and Kaufman, 1965, 668). It is conceivable that the office of City Administrator or Deputy Mayor could meet this need by development towards substantive powers, precisely defined by custom, as well as by charter, and bounded by overall responsibility to the Mayor. However, good Mayors are jealous of their position and powers —and rightly so. Mayor Clark liked to run the show himself; Mayor Daley vetoed a Charter Commission recommendation for a Chief Administrative Officer.

The administrator-entrepreneur

Officials of this kind have derived their position from formal charter revisions and in the main have not been able to build themselves more power than the constitution permitted. City political systems have nevertheless given scope for a type of administrator-entrepreneur who has had a profound impact on city government. Among examples of the type, Robert Moses of New York represents, in a grossly exaggerated form, the administrator who builds an empire based on authority in independent boards, and at times in defiance of the Mayor. Edward Logue represents a more common type, the development administrator accorded full power to carry out wide-ranging functions. Both illustrate the possibilities of accumulating power, and the success that sometimes rewards the official who 'thinks big', organizes a crusade, adopts community rather than specialist goals, and does not neglect public relations. Moses added to these qualities a concern for independent power which makes him a splendid companion portrait to the British exemplar of the same phenomenon, Lord Reith, creator of the British Broadcasting Corporation and the concept of public service broadcasting. (Chadwick or Morant might conceivably stand in the same gallery, and Moses himself would include Baron Haussman.)

(i) Robert Moses

Robert Moses' great skill lay in the accumulation of appointive offices and their exploitation for creative municipal purposes. He had no taste for elective office and was defeated in his one attempt to gain a popular majority, for the Governorship of New York State in 1934. But at one time he held ten municipal and state offices, including the Chairmanship of the Triborough Bridge and Tunnel Authority, Parks Commissioner and City Construction Co-ordinator. He drew salary for only two of these posts: basically he was interested in constructive power, not wealth.

He was both brilliant and unprincipled in his pursuit of power. He exploited his contacts in the State Capital. As Parks Commissioner he was given a special and unwarranted status in the Budget Bureau and secured appointment, by La Guardia, to the City Planning Commission, despite an apparent conflict of interest, and, more seriously, despite his opposition to the concept of disciplined and comprehensive planning. His part in the epic struggle over the building and control of the City's airports (narrated in Stein, 1952, 143-197) seems chiefly concerned with his own power in the ultimate administrative structure. Moses probably understood the immense financial and administrative problems as well as anybody, but the attempt to create and maintain a Moses-dominated City Airport Authority, and the long drawn out rivalry with New Jersey, only postponed Mayor O'Dwyer's final and logical solution of handing the airports to the bi-state Port of New York Authority. Like all strong personalities, Moses had admirers and detractors. The judicious and authoritative Sayre and Kaufman write: 'His many positions are as much a result of his effectiveness as a cause of it; work gravitates to him because of his unique combination of talents. He is not only brilliant, articulate, imaginative, energetic, honest and confident, but he is completely unawed by the stature of those he deals with' (Sayre and Kaufman, 1965, 341-2). His drive and energy could of course be interpreted less favourably as impatience, ruthlessness, a stubborn and egocentric judgement. His disregard for what others thought of him (Moses 'doesn't give a damn') amounted sometimes to contempt: well-meaning and able people were dismissed as 'long-haired' (planners); 'socialists' (New Dealers, though Moses himself was happy to exploit the New Deal's financial subsidies); 'crackpots' (almost anybody).

Yet his success, both in getting power and using it constructively, cannot be denied. Much of this must be ascribed to sheer talent, energy and dedication. He was educated at Columbia and Oxford Universities, and wrote a good (and highly favourable) book on the British civil service. But intellectual power was refined by political skills of a high order. In particular Moses took great care over the presentation of his plans, preparing elaborate brochures, often in advance of approval, and feeding material to well-disposed publicists. He was not troubled by self-doubt and was sustained by an aggressive sense of righteousness and mission, which was as strong as Reith's, though it could not claim an origin in the manse. However, talent and drive are never enough for such success as Moses enjoyed. His success owed something too, to the support of able colleagues, and to the backing of two powerful Mayors, La Guardia and O'Dwyer. The *Times* proved a consistent champion, and the people of New York could see, use and enjoy the great public works associated with the name of Robert Moses.

These, the Triborough Bridge, the parks and parkways, are indeed a monument to his achievement, which spanned a period of more than thirty years of public office, and was maintained in a highly political city under five Mayors and three parties. But if the fruit of Moses was abundant, the costs too were not negligible. Moses did not believe in comprehensive planning and helped to destroy its early development in New York. His ideas on slum clearance and urban renewal were old-fashioned: he was too little concerned for the problems of relocation, density, segregation and private exploitation; he cared little for public scrutiny, aesthetic townscape, and long-range planning. Critics were curtly dismissed: 'people always stand in the way of progress' (see Lowe, 1967, 45-109 passim). Lowe admits that, 'No one else in American history produced so much local public improvement and urban reconstruction—about $5,000,000,000 by 1960.' But, she adds, 'no single person contributed more through his works and his methods to New York City's problems—the remote, fractionalized local government; the unplanned private overgrowth; the traffic congestion; the inhumanity and citizen discontents; the real estate "project" approach to community building; and the abdication of political and business leadership' (Lowe, 1967, 47-8).

For all the criticism, no city has yet solved the problems of urban renewal, and New York, without Moses, has not enjoyed spectacular success in its public construction programmes. Like many great creative men in public life, Moses found a vacuum and left one. Like other great creators, too, his achievement depended on the matching of his talents with the times. Moses' contribution was to recognize the moment for great public works to meet the needs of the expanding city. A generation later both the moment and the needs were radically different. But a city, like a nation, ought to content itself perhaps with a temporary and time-bound burst of creativity.

(ii) Edward Logue

Edward Logue has also been a successful administrator, a Moses without the warts. Logue was New Haven's Development Administrator under Mayor Lee; moved to Boston as Redevelopment Administrator under a distinguished urbanist, Mayor John Collins, now a Professor at MIT, and then to New York State as executive officer of the State's Urban Redevelopment Corporation (whose writ runs even into New York City). Where Moses stayed in one place and accumulated offices, Logue has moved to situations in which his creative drives could be employed without inhibition. Like Moses he gave detailed attention to the preparation and presentation of plans, developing his programme in secret, conciliating in advance, and unveiling a finished design in a way which ensured approval and discouraged even minor amendment (Polsby, 1963, 73-4, and Lowe, 434). Like Moses, too, he has had the good fortune to be supported by Mayors, both in the city, and in tapping federal resources. In particular Logue's work in New Haven cannot be separated from that of Mayor Richard Lee. In Boston, he had the advantage of formal authority over both urban renewal and city planning. Logue's plans, though large and ambitious, have been carefully designed. He had a conception of a whole community, requiring schools and other amenities, as well as housing. He was aware of the problems of securing profitable private development, and maximum federal assistance and was a skilful manipulator of the financial side of renewal. He was sensitive both to aesthetic considerations, and to the needs and feelings of the people who are to be housed or rehoused. He had faith in 'the utility of bricks and mortar, not only in improving physical conditions, but also in renewing the life of a city' (Logue quoted in Lowe, 1967, 551).

Robert Moses was a unique version of the administrator-entrepreneur; but Edward Logue is one of a recognizable type of urban administrator; Bacon in Philadelphia, Danzig in Newark are comparable figures. Their achievements are not likely, however, to be continuous, widespread and spectacular. For this, the problems are too difficult, the political and administrative resources inadequate. Mayor Lindsay has had the benefit of Logue's advice and the assistance of some of his staff, but New York's problems have not yielded as easily as New Haven's (and New Haven itself suffered serious rioting in August 1967). One of Logue's ablest lieutenants from his New Haven staff was specially recruited by Lindsay to head New York's Human Resources Administration. He resigned within a year. But effective or not, the type and style of crusading Administrator-entrepreneur is indicated in the very name—Human Resources Administrator.

5
Conclusion

The role of political institutions and executive authority

A characteristic feature of modern political science is to deny the assumption that the political system has purposefulness and rationally directed power. 'Perceptions and decisions about public needs are made by highly diverse and segregated power centres, each operating with little relation to, or knowledge of, the other' (Wood, 1963, 109). Norton Long describes metropolitan politics thus: 'Issues and problems have a career and over time processes of interaction develop through which interested and powerful parties exercise influence over the outcome' (Long, 1962, 157). Local politics, lacking some of the features of national politics which make for integration, suffers even more from fragmentation.

Thus the urban political system is perceived as diffuse and fragmented in a random manner, a mere field of force in which the most substantial function of the executive authority is brokerage. This kind of analysis has not been wholly free of values in that there is a tendency to exaggerate the equitability of pluralism, as well as its freedom, and to stress the virtues of system maintenance and stability as if the alternative were the instability of a banana republic.

The tendency of this account is to re-emphasize the

role of political institutions and executive authority. The justification for this must lie ultimately in the coherence and credibility of the analysis of city politics to which it contributes. But there are specific sources of support for such an interpretation. First, it is clear that there is no fundamental incompatibility with other interpretations. The political resources of city government for competitive struggle are not negligible. Brokerage leadership is still leadership; choice between competing groups may still be, within those limits, free choice; bargaining includes persuasion short of coercion and, given the inequalities of bargaining power, may favour the executive authority. Second, it is conceivable that the power of local authorities varies over time and that it has been moving in the 1960's towards greater power. There are obvious drives in this direction; the urgency of urban problems is producing both a problem-oriented leadership and citizenry. The new prominence of federal welfare programmes suggests that in the bigger cities at least, old law enforcement and even new 'law and order' are not adequate objectives to meet the demands made of the system. The exploitation of the media by leading politicians may also affect outcomes.

Finally, this view of the importance of executive authority is neither new nor individual, nor singular. It is to be found in many other accounts and appears as a centralizing tendency arising from the middle class ethos in Banfield and Wilson (1963); and is supported by some recent research on policy outcomes (e.g. Wolfinger and Greenstein, 1969). It does not amount to a claim that executive authority is exercised in a power-vacuum, nor that cities are or have become manageable enterprises. (The achievements of even the most successful strong Mayors are limited, and some have had little success.) No myths are exploded and no chairs rock. The urban political system is still seen as in Norton Long's phrase 'an ecology of games'. It is, indeed, rather like a continuing series of games of tennis. But the players move from court to court, playing in more than one game. Some games have umpires, audiences, and a few have full television coverage. Some are played as a competitive ritual, with costumes but without a ball. The tennis analogy admits the possibility of a struggle for mastery in which victory is built up piecemeal in game, set and match, and in which some players and some teams show a persistent tendency to win certain kind of games. But unlike Wimbledon or Forest Hills, there are few big prizes and even fewer champions.

Crisis and stability in the urban political system

'Crisis' is a journalistist's not a scholar's term, and it is easy to show that the situation of the cities is not critical in the sense of imminent disaster. The apocalyptic vision of the suburbs promoted by some urbanists is plainly no cause of anxiety to most suburban dwellers. 'The American urban place is a non-city because Americans wish it to be just that' (Prof. Elazar in Federal Role, 1966, 13, 2703). The rather more serious problems of urban poverty and the ghetto are more difficult to dismiss, though it is well to remember that poverty and race are still substantial problems in rural areas too.

In the cities there is indeed evidence in new Negro political organization and militant activity that the Negroes are more than just another pressure group to be accommodated. The system is likely to be under strong pressure at the very least to alleviate the condition of the urban Negro by welfare outputs. But the system was not designed for this: its grass roots, parochialism, its conservatism, its preoccupation with taxes and law enforcement, its middle class ethos, its visibility and vulnerability, all combine to inhibit the production of positive and controversial policies: the record in fluoridation and desegregation supports this. Thus the city's problems may be beyond the reach of politics. There are two possible solutions. Cities may be compelled to respond by the external pressures of federal and state governments. Second, and perhaps following on this, executive power will be developed, based on a coalition of groups, or a coherent party organization. Given the conditions of American city politics, such developments would tend to lapse when the immediate threat to the system was past.

Comparative aspects

British cities are similar to American cities in many of the ways which affect politics. In both countries about three-fifths of the population live in cities larger than 100,000. The big cities are spread out and split between centre and suburb, and with boundaries ill-adjusted to the service of communities. The physical plant of the city is decayed and some of its citizens afflicted by poverty and other social deprivations. In consequence, the municipality is pressed to undertake urban renewal and welfare services. At the same time, the citizens are generally apathetic towards municipal politics, but liable to occasional public anger if hurt or affronted (in both countries, housing and education are frequent sources of protest).

There are two major historical and social differences which affect urban politics. Britain has no historical experience equivalent to immigration and territorial expansion, and the size and concentration of the coloured population makes colour a serious local problem only in a few cities. Britain has, on the other hand, a nationalities problem, which is likely to affect the pattern of central-local power.

The present centralization of power in Britain is in fact the most remarkable political difference in the local government of the two countries. Although American cities are seriously constrained both by their federal ties and their own structural weaknesses, it is still possible for the American city to develop and deploy political power beyond that of a British municipality.

In America the power available to the system as a whole may be greater than in Britain, but is subject to a debilitating diffusion within the system. Power is fought for and shared by pressure groups in the community, and by competing administrators and politicians within the institutional framework. The result is often stalemate or dispersion. It may also happen, however, that the struggle for power gives rise to a dynamism in politics, and produces greater concentrations of power and greater problem-solving and programme potential. Thus, while it is not true that American city government is consistently more powerful than the British, the potential, the upper limits of power which a government may develop, are very much greater. The system lacks the consistency of Britain's limited grants of power, but generates more energy in seeking and assembling power.

There is, in consequence, more political activity in American cities than in British cities. Pressure groups are more active as well as more open. The operations of political parties, whether in competitive or dominance situations, are characterized by the use of patronage, which has sharper effects than the policy disciplines of a local British party. There are more signs in American cities of reformist groups and amateur insurgency. The number and comparative vigour of some civic groups quite outshine the activities of Britain's rate-payers, tenants and parents' associations, and civic societies. Britain lacks both the open, participant tradition of American politics and the specific challenge of referenda and elective ad-hoc boards. In the face of so much political activity, American cities necessarily pursue a politics of accommodation and bargaining, leavened by political or administrative leadership. The result is a system which ranges from agitation and inertia to dynamism and power.

The study of city politics

The study of American city politics suggests areas and processes for further investigation in Britain. These include:

the political effects of non-partisanship;

a search for an equivalent of the boss and patronage, i.e. the mechanisms of party control;

a delineation of local political ethos, in terms of reform, or in the bleaker terms of the Ratepayers' Associations;

an explanation of religious, linguistic, moral and nationalist cleavages, especially in those forgotten non-English parts of the United Kingdom;

and of pressure groups, which though more discreet, less open than in USA, operate nevertheless and perhaps more effectively;

a consideration of the political consequences of regional (i.e. metropolitan) reorganization in terms of parties, groups and people;

the development of typologies of British cities based on community structure, politics and policy outputs.

Such studies might lead to a better understanding of the nature of city politics, and to an elucidation of the basis on which recommendations might be made. It should then be possible to examine critically and to assess the impression of one English observer of American cities, that the good government of British cities requires greater political activity, including the obligation to win the support of voters more frequently; wider scope for executive leadership; and the re-introduction of the 'ad-hockery' which was once associated with bursts of creative if ill-co-ordinated energy.

Suggestions for further reading

The following is a short course of reading to lead into the study of American city politics. For bibliographical details, see Bibliography.

i Adrian, 1961—a general text-book.
ii Banfield and Wilson, 1963—a basic interpretive text.
iii Four short pieces: Brogan, 1954, chapter 4; Dahl, 1967; Wood, 1963; Jacob and Lipsky, 1968.
iv Readers edited by Banfield, 1961; and Williams and Press, 1961.
v Three important monographs: Dahl, 1961; Banfield, 1961; Williams and Adrian, 1963.
vi Kaplan, 1963—urban renewal as a case study.
vii Wilson, 1968—a selection of recent work on the correlations of policy outputs.
viii R. Alford in Schnore and Fagin, 1967: a comparative review of four major works.

Select bibliography

ADRIAN, C. R. (1961), *Governing Urban America*, New York: McGraw-Hill Book Company.

ADRIAN, C. R. (1958), 'Leadership and decision making in manager cities, a study of three communities', *Public Administration Review*, 18, 3.

AGGER, R. E., GOLDRICH, D. and SWANSON, B. E. (1964), *The Rulers and the Ruled*, New York: John Wiley.

ALFORD, R. and LEE, E. C. (1968), Voting Turnout in American cities, *American Political Science Review*, LXII, 3.

ALFORD, R. and SCOBLE, H. (1965), 'Political and socio-economic characteristics of American cities', *Municipal Yearbook*, 82-97.

ALMOND, G. and VERBA, S. (1963), *The Civic Culture*, Princeton University Press.

BANFIELD, E. C. (1961), *Political Influence*, New York: The Free Press of Glencoe.

BANFIELD, E. C. (1961), *Urban Government, a Reader in Administration and Politics*, New York: Free Press of Glencoe. Revised edition, 1969.

BANFIELD, E. C. (1965), *Big City Politics: a comparative guide*, New York: Random House.

BANFIELD, E. C. and WILSON, J. Q. (1963), *City Politics*, Cambridge, Mass.: Harvard University Press and MIT Press.

BOOTH, D. A. (1963), *Metropolitics: The Nashville Consolidation*, East Lansing, Michigan: Institute for Community Development and Services.

BROGAN, D. (1954), *An Introduction to American Politics*, Hamish Hamilton.

BRYCE, JAMES (1888), *The American Commonwealth*, in edition of 1959 edited by L. Hacker, Capricorn Books, New York: Putman.

CAMPBELL, A. and BURKHEAD, J. (1968), in PERLOFF, H. and WINGO, L., *Issues in Urban Economics*, Baltimore: John Hopkins Press.

CARALEY, D. (1966), *New York City's Deputy Mayor City Administrator*, New York: Citizens Budget Commission Inc.

CARRELL, J. J. (1962), *The Role of the City Manager*, Kansas City, Mo: Community Studies Inc.

CURLEY, J. M. (1957), *I'd do it again*, Englewood Cliffs, N.J.: Prentice Hall.

DAHL, R. (1961), *Who governs?*, New Haven: Yale University Press.

DAHL, R. (1967), 'The City in the future of democracy', *American Political Science Review*, LXI, 4.

DAHL, R. (1967), *Pluralist democracy in the US*, Chicago: Rand McNally.

DUNCAN, O. D. (1957), 'The Optimum size of cities' in HATT, P. K. and REISS, A. J. (ed.) *Cities and society*, Glencoe, Illinois: Free Press.

ELAZAR, D. J. (1967), 'Urban problems and the federal government', *Political Science Quarterly*, 82.

EYESTONE, R. and EULAU, H. (1968), 'City Councils and Policy Outcomes: Development Profiles', in Wilson, 1968.

FROST, R. T. (1961), 'Stability and change in local politics', *Public Opinion Quarterly*, 25.

GANS, HERBERT J. (1962), *The Urban Villagers*, New York: Free Press of Glencoe.

GARRETT, C. (1961), *The La Guardia Years: Machine and Reform Politics in New York City*, New Brunswick, N.J.: Rutgers University Press.

GLAAB, C. N. and BROWN, A. T. (1967), *A History of Urban America*, New York: The Macmillan Company.

GOODALL, L. (1968), *The American Metropolis*, Columbus, Ohio: C. E. Merrill.

GOSNELL, HAROLD (1968), *Machine Politics, Chicago Model*, second edition with Foreword by T. J. Lowi, University of Chicago Press.

GRAY, K. E. and GREENSTONE, D. (1961), 'Organized Labour in City Politics' in Banfield Reader, 368-379.

GREENSTONE, J. D. and PETERSON, P. E. (1968), 'Reformers, Machines, and the War on Poverty', in Wilson, 1968.

GREER, SCOTT (1963), *Metropolitics*, New York: John Wiley & Sons.

GREER, SCOTT (1965), *Urban Renewal and American Cities*, Indianapolis: Bobbs-Merrill.

HARRELL, C. A. and WEIDFORD, D. C. (1959), 'The City Manager and the Policy Process', *Public Administration Review*.

JACOB, H. and LIPSKY, M. (1968), in IRISH, M. *Political Science*, Englewood Cliffs, N. J.: Prentice-Hall.

KAPLAN, H. (1963), *Urban Renewal Politics: Slum Clearance in Newark*, New York: Columbia University Press.

KAPLAN, H. (1967), *Urban Political Systems: a functional analysis of Metro Toronto*, New York: Columbia University Press.

KESSEL, J. (1962), Governmental structure and political environment, *American Political Science Review*, LVI, 4.

LEE, E. C. (1960), *The politics of non-partisanship*, Berkeley: University of California Press.

LERNER, MAX (1957), *America as a civilization*, Jonathan Cape.

LEVIN, M. B. (1965), *The Alienated Voter: Politics in Boston*, New York: Holt, Rinehart and Winston.

LINEBERRY, R. and FOWLER, E. (1967), 'Reformism and

public policies in American Cities', *American Political Science Review*, LXI, 3.

LOCKARD, DUANE (1963), *The Politics of State and Local Government*, New York: Macmillan.

LONG, N. (1962), *The Polity*, Chicago: Rand McNally.

LOWE, JEANNE R. (1967), *Cities in a race with time*, New York: Random House.

LOWI, T. J. (1964), *At the pleasure of the Mayor*, New York: Free Press of Glencoe.

MCCORKLE, S. A. (1952), 'What the Council expects from the City Manager', *Public Management*, July.

MEYERSON, M. and BANFIELD, E. C. (1964), *Politics, Planning and the Public Interest*, New York: The Free Press of Glencoe.

MORLAN, R. L. (1960), *Capitol, Courthouse and City Hall*, Boston: Houghton Mifflin.

MUNICIPAL YEARBOOK (annually), International City Managers Association, Washington, DC.

POLSBY, N. W. (1963), *Community Power and Political Theory*, New Haven: Yale University Press.

REICHLEY, J. (1959), *The Art of Government*, New York: The Fund for the Republic.

RIDLEY, C. (1958), 'The Role of the City Manager in Policy Formulation', *Public Administration Review*, 18.

ROSSI, P. H. and DENTLER, R. A. (1961), *The Politics of Urban Renewal*, New York: Free Press of Glencoe.

SALISBURY, R. H. and BLACK, G. (1963), 'Class and party in Partisan and Non-Partisan Elections', *American Politrical Science Review*, LVII, 3.

SAYRE, W. S. and KAUFMAN, H. (1965), *Governing New York City: Politics in the Metropolis*, New York: W. W. Norton.

SCHNORE, L. and FAGIN, H. (1967), *Urban research and policy planning*, Beverly Hills, California: Sage Publications.

STEIN, H. (ed.) (1952), *Public Administration and Policy*

Development, New York: Harcourt, Brace.

STONE, H. et al. (1940), *City Manager Government in the US, A review after 25 years*, Chicago: Public Administration Service.

STONE, H. et al. (1940), *City Manager Government in nine cities*, Chicago: Public Administration Service.

STRAETZ, R. A. (1958), *PR Politics in Cincinnati*, New York: New York University Press.

TAX FOUNDATION (1966), *Allocation of the Federal Tax Burden by State*, New York.

WILDAVSKY, AARON (1964), *Leadership in a small town*, New Jersey: The Bedminster Press.

WILLIAMS, O. and ADRIAN, C. R. (1963), *Four Cities*, Philadelphia: University of Pennsylvania Press.

WILLIAMS, O. P. and PRESS, C. (1961), *Democracy in Urban America: Readings on Government and Politics*, Chicago: Rand McNally.

WILLIAMS, O. P. et al. (1965), *Suburban differences and metropolitan conflicts*, Philadelphia: University of Pennsylvania Press.

WILSON, J. Q. (1962), *The Amateur Democrat: Club Politics in Three Cities*, University of Chicago Press.

WILSON, J. Q. (ed.) (1968), *City Politics and Public Policy*, New York: John Wiley.

WILSON, J. Q. and BANFIELD, E. C. (1964), 'Public regardingness as a value premise in voting behaviour', *American Political Science Review*, LXIII, 4.

WOLFINGER, R. and FIELD, J. O. (1966), 'Political Ethos and the Structure of City Government', *American Political Science Review*, LX, 2.

WOLFINGER, R. and GREENSTEIN, F. (1969), 'Comparing Political Regions: the case of California', *American Political Science Review*, LXIII, 1.

WOOD, ROBERT C. (1958), *Suburbia: its people and their politics*, Boston: Houghton Mifflin.

WOOD, R. C. (1961), *1,400 Governments*, Cambridge, Mass.:

Harvard University Press.

WOOD, R. C. (1963), 'The contributions of political science to urban form' in HIRSCH, W. Z. (ed.) *Urban Life and Form*, New York: Holt Rinehart and Winston.

United States Papers

Creative Federalism (1967), Hearings before the Subcommittee on Intergovernmental Relations of the Committee on Government Operations of the US Senate.

Federal Role (1966 or 1967), Federal Role in Urban Affairs, Hearings before the Subcommittee on Executive Reorganization of the Committee on Government Operations, US Senate.

Kerner Report (1968), Report of the National Advisory Commission on Civil Disorders (the edition cited is by Bantam Books, New York).

Kestnbaum Report (1955), Report of the Commission on Intergovernmental Relations.

Revenue Sharing (1967), Revenue Sharing and its Alternatives: What future for Fiscal Federalism?, 3 volumes, Subcommittee on Fiscal Policy of the Joint Economic Committee, US Congress.

Urban America: Goals and Problems (1967), Subcommittee on Urban Affairs of the Joint Economic Committee.

Also:

Hearings of the subcommittees on Independent Offices and HUD of the Senate Appropriations Committee; and on Housing and Urban Affairs of the Senate Committee on Banking and Currency.